A Nature Company Guide

NATURAL GARDENING

A Nature Company Guide

NATURAL GARDENING

JOHN KADEL BORING, ERICA GLASENER,
GLENN KEATOR, JIM KNOPF,
JANE SCOTT, SALLY WASOWSKI

CONSULTANT EDITOR
RG TURNER JR

THE NATURE COMPANY

TIME LIFE BOOKS

AG

A Nature Company Guide published by Time-Life Books

Conceived and produced by Weldon Owen Pty Limited
43 Victoria Street, McMahons Point, NSW, 2060, Australia
A member of the Weldon Owen Group of Companies
Sydney • San Francisco • London
Copyright © 1995 Weldon Owen Pty Limited

The Nature Company owes its vision to the world's great naturalists:
Charles Darwin, Henry David Thoreau, John Muir, David Brower,
Rachel Carson, Jacques Cousteau, and many others.
Through their inspiration, we are dedicated to providing products and
experiences which encourage the joyous observation, understanding, and
appreciation of nature. We do not advocate, and will not allow to be sold in
our stores, any products that result from the killing of wild animals for trophy
purposes. Seashells, butterflies, furs, and mounted animal specimens fall into
this category. Our goal is to provide you with products, insights, and
experiences which kindle your own sense of wonder and help you to feel
good about the world in which you live.
For a copy of The Nature Company mail-order catalog, or to learn the
location of the store nearest you, please call 1-800-227-1114.

THE NATURE COMPANY
Catherine Kouts, Steve Manning, Sydni Scott,
Tracy Fortini, Terry McGinley

TIME-LIFE BOOKS
Time-Life Books is a division of Time Life Inc.,
a wholly owned subsidiary of THE TIME INC. BOOK COMPANY

VICE PRESIDENT AND PUBLISHER: Terry Newell
EDITORIAL DIRECTOR: Donia A. Steele
DIRECTOR OF NEW PRODUCT DEVELOPMENT: Regina Hall
DIRECTOR OF SALES: Neil Levin
DIRECTOR OF CUSTOM PUBLISHING: Frances C. Mangan
DIRECTOR OF FINANCIAL OPERATIONS: J. Brian Birky

THE NATURE COMPANY GUIDES
PUBLISHER: Sheena Coupe
MANAGING EDITOR: Lynn Humphries
PROJECT EDITORS: Julia Cain, Dawn Titmus
EDITORIAL ASSISTANTS: Greg Hassall, Vesna Radojcic
SUB EDITOR: Glenda Downing
ART DIRECTOR: Hilda Mendham
DESIGNERS: Stephanie Cannon, Gabrielle Tydd
JACKET DESIGN: John Bull
PICTURE RESEARCH: Connie Komack, Pictures & Words,
Rockport, Massachusetts; Gillian Manning
PRODUCTION DIRECTOR: Mick Bagnato
PRODUCTION COORDINATOR: Simone Perryman
VICE-PRESIDENT INTERNATIONAL SALES: Stuart Laurence
COEDITIONS DIRECTOR: Derek Barton

Library of Congress Cataloging–in–Publication Data
Natural Gardening/John Kadel Boring … [et al.];
consultant editor, R.G. Turner, Jr.
    p. cm. — (A Nature Company guide)
    Includes bibliographical references (p.) and index.
    ISBN 0–7835–4750–1 (trade)
    1. Natural landscaping. 2. Gardening to attract wildlife. 3. Natural
landscaping—United States. 4.Gardening to attract wildlife—United States.
I. Boring, John Kadel. II. Turner, R.G. III. Series.
SB439.N4 1995   95–1225
635.9'5173—dc20   CIP

Manufactured by Kyodo Printing Co. (S'pore) Pte Ltd
Printed in Singapore

A Weldon Owen Production

Nature—wild nature—dwells in gardens just as she dwells in the tangled woods, in the deeps of the sea, and on the heights of the mountains; and the wilder the garden, the more you will see of her there.

Adventures in Green Places,
Herbert Ravenel Sass (1884–1958), American nature writer

CONTENTS

———— ✦ ————

FOREWORD

Rachel Carson once said the way to truly appreciate your surroundings is to "ask yourself, 'What if I had never seen this before? What if I knew I would never see it again?'" Traveling across the continent, walking through a local park, or exploring the wilds of your own garden—wherever you find yourself—it's important to slow down and allow yourself to notice things.

John Muir recommended walking as the best way to get in touch with nature and yourself. Seeking a new perspective, he walked all the way from Wisconsin to Florida. But you needn't go so far to kindle the spirit of exploration in your life. Thoreau coaxed a lifetime of discovery out of the woods surrounding a thimble-sized pond.

I've traveled across North America, from the lowlands of the Everglades to the peaks of Yosemite, and I carry the sights, sounds, and scents of those places in my mind like treasure. But I still have many of my best adventures in my own garden. I've noticed that within the cycles that provide rhythm to the year—rainy seasons and dry, bloom and harvest—there are infinite variations that make each year unique. Every time I venture out, I am rewarded by a glimpse of new growth, a bird I've never seen there, a bright jewel of a beetle, a spider web glistening in the sun.

That's what *The Nature Company Guides* are all about: awakening a sense of adventure and delight in our everyday lives. We hope they will inspire you to close the book and open the door to the world of nature.

PRISCILLA WRUBEL
Founder, The Nature Company

INTRODUCTION

In an age of environmental crisis, people tend to lose sight of the fact that not everything we do to the land is bad for the plants and animals with whom we share it. Most of us automatically assume that the best we can do in nature is the least: to leave the land alone. Yet as many a gardener has observed, it is possible to make changes in the land that actually contribute to its ecological health—that add to the abundance and diversity of life in a place.

Natural gardening is the name of a method, and a rapidly growing movement, that aims at enhancing the ecological well-being of the places where we live. This can mean a great many things: overthrowing the tyranny of the American lawn in your frontyard; planting a wildflower meadow; digging a pond; providing flora that will attract the local fauna; restoring native plants to a place from which they've vanished.

As the natural gardener soon appreciates, one needn't travel to the wilderness in order to observe the workings of a rich and vital ecosystem. In fact, you hold in your hands a tool with which to establish such an ecosystem in your own backyard. Part instruction manual and part field guide, *Natural Gardening* will show you, step by step, how to improve the ecology of your particular corner of the Earth—to make it more alive. What could be more worthwhile than that?

MICHAEL POLLAN
Author of Second Nature: A Gardener's Education

Through intimate association with the living things around us, we reach out beyond the narrow human sphere into the larger natural world …

ALEXANDER FRANK SKUTCH (b. 1904),
American naturalist

CHAPTER ONE

CREATING *a* NATURAL GARDEN

GARDENING *with* NATURE

A natural garden is a place that is attractive and welcoming for both people and wildlife.

Beauty, discovery, wonder, and joy—the possibilities that lie in a backyard, a frontyard, a sideyard, or even a schoolyard cannot be underestimated. Gardening with nature is an approach that opens up innumerable opportunities and creates myriad possibilities. It can turn a disappointing site into a delightful experience, involve food production and wildlife, and even an appreciation of other cultures. Gardening with nature is a means of connecting people and their environment. It's about working with the natural climate—the sun and the shade, the soils and the topography—and the local wildlife. It is a wonder-filled way to make the very most of any location.

By contrast, most modern American landscapes are models of like-minded conformity, and are built upon a heritage of dominating nature rather than learning from it and enjoying working with it. The destruction of the native vegetation has left a society with little contact with the natural environment. Widespread loss of habitat has put many native species, both flora and fauna, at risk of extinction. The use of toxic chemicals has resulted in serious degradation of land, water, and atmosphere.

CHANGING ATTITUDES

As we approach the end of the twentieth century, new attitudes about stewarding our land have begun to take hold throughout North America. The organic gardening movement has grown steadily; now, even organically grown farm produce has achieved mass-market recognition. Increasingly, home gardeners are searching out nontoxic methods of pest control, and taking the time to learn about the benefits of composting. There is a resurgence of interest in regional designs built upon natural models.

THE NATURAL MODEL

A garden is, by its nature, artificial, yet it contains natural elements. To use nature as the model for the garden is to recognize the incredible fit that native plants, and animals, have to the land. Gardens based on

A NEEDLEPOINT DEPICTION *(top) of early European settlers enjoying their natural surroundings. Plant vibrant perennials, such as daisies and yarrows (right), to bring your garden to life.*

SHRUBS PROVIDE *food and protection (above) for garden visitors such as the colorful northern cardinal. It is rewarding to work with nature (left) and create a beautiful natural garden.*

such a model are kinder to the environment and less wasteful of precious resources. They are easier to maintain, because native plants are generally well suited to the local conditions and may be more resistant to pests and diseases. Knowledge of the native flora gives insights into the appropriateness of plants introduced from other areas.

Plantings to attract wildlife can provide homes for birds, mammals, reptiles, and amphibians, as well as bees, butterflies, and other species of beneficial insect. Gardens that are planted with seasonal change in mind can be more visually interesting and varied throughout the year for the gardener and more appealing for desirable wildlife, while discouraging wildlife problems.

The quest to put ourselves at ease with the natural world

LOW-MAINTENANCE, *natural desert gardens (right) offer much to wildlife, from edible cactus fruits to palo verde trees for shade and nesting materials.*

is constant. Searching for the essence of original, natural landscapes is a relevant and rewarding part of this quest. *Natural Gardening* is about this satisfying search for natural landscapes and the fascinating world of wildlife that can be observed in the natural garden. The emphasis throughout the book is on planting to attract wildlife as a rewarding way to open vistas into the natural world around us.

JUST THE START

Natural Gardening should not be the only gardening book you own; as an introduction to the subject, it is designed to focus on the principles more than the practice of natural

gardening. Many other books (see Further Reading, pp. 274–5) will present more completely the process of designing an attractive and functional garden, while others discuss in detail the day-to-day cultural practices necessary for planting and maintaining a garden. Still others offer a more detailed compendium of plants, both native and otherwise, suitable for growing in each region of the country.

Natural Gardening is an invitation to see that gardening can offer the most rewards when it is done in concert with nature, rather than in competition with it. Gardens are best when they are designed and created with inspiration from nature, rather than by radically altering nature. And, no matter the size of the garden, gardening with nature is fun, beautiful, relevant, and wonder-filled.

THE NATURE *of* PLANTS

*Plants set the stage in a natural garden, weather changes the scenes,
wildlife form the cast, and people are the interactive audience.*

Plants, in their glorious variety, form the basis of gardening. From the daintiest mosses to towering trees, they all serve wildlife in some manner. Bark, twigs, leaves, flowers, fruits, and seeds provide nourishment; cavities and branches form nesting sites; and foliage provides valuable cover from predators and harsh weather. Even plants that appear inhospitable to humans can, in fact, offer shelter. Gila woodpeckers, for example, raise their broods in the fleshy trunks of the huge saguaro cactus (*Carnegiea gigantea*) of the Desert South-west. These nesting sites protect them from the harsh extremes of the desert climate, and the thorns deter predators.

THE PLANT KINGDOM

The world of plants is broadly divided into vascular and non-vascular plants. The more advanced vascular plants have specialized water- and food-conducting tissues. These enable the plants to grow in a greater range of forms and become much larger—and, therefore, are of greater value to gardeners—than those that are nonvascular (such as algae, lichens, and mosses).

Vascular plants reproduce either by spores (the ferns and their allies) or by seeds. Plants in both groups make valuable garden plants, and are classified scientifically by the structure of their reproductive parts. The oldest seed-bearing plants are called the gymnosperms; though the name means "naked seed", seeds of these plants are usually held in woody or fleshy cones. The best-known gymnosperms are the cone-bearing pines and firs, and the fleshy-fruited ginkgo, juniper, and yews. The angiosperms, or flowering plants, form the largest and most diverse group in the plant kingdom, with more than 250,000 species, ranging from grasses to trees, and cacti to waterlilies. Flowering plants produce seeds enclosed in fleshy, papery, or woody fruits to protect the fragile seed and to allow it to disperse away from the parent plant.

Herbaceous plants sometimes die to the ground in freezing weather or in drought, and have stems that are usually green and soft-tissued. Woody plants develop permanent

GARDEN VARIETY *Plant your garden with a range of annuals, perennials, evergreens, and deciduous plants (below and right) to create different habitats and for an ever-changing garden scene throughout the year.*

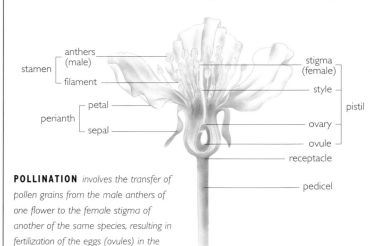

stamen — anthers (male) / filament
perianth — petal / sepal
stigma (female)
style
pistil
ovary
ovule
receptacle
pedicel

POLLINATION *involves the transfer of pollen grains from the male anthers of one flower to the female stigma of another of the same species, resulting in fertilization of the eggs (ovules) in the ovary and the production of seeds.*

SOME PLANTS RELY ON THE
wind to disperse their seed (left).
Encasing seeds in tasty fruits, such
as berries (above) and nuts, is a
strategy that depends on animals,
especially birds, for dispersal.

stems, or trunks, of cellulose
tissue toughened with lignin.

POLLINATION PLOYS

Plants, like animals, depend
upon the uniting of males and
females to produce offspring.
To circumvent their inability
to move around, plants have
developed an array of strategies
to ensure pollination. The
gymnosperms, grasses, and
many forest trees rely upon
the wind, releasing great
masses of pollen to increase

the chances for some to
actually land on female
flowers of the same species.

Many plants, however, use
their flowers to attract visiting
animals to do the job. Bright
colors, fragrance, patterns, or
shapes that mimic sexual
partners entice creatures such
as hummingbirds, butterflies,
and bees to move pollen from
male to female flower parts.
The reward for the visitor is
nourishing pollen, nectar,
or flower tissue.

SEED DISPERSAL

For a plant to reproduce
successfully, some of its seeds
must fall on fertile ground,
germinate, and in time,
reproduce. Ideally, the seeds
will land where they will not
compete with the parent plant
for water, sunlight, and nutrients. Seeds are dispersed in a
variety of ways, including
being carried by wind and
animals. Some seeds cling to
passing creatures, such as the
burdock seeds that inspired
the invention of Velcro.

LIFE CYCLES

In horticulture, grasses and
herbaceous garden flowers can
be grouped according to their
life cycle. Annuals germinate,
flower, produce seed, and die
all in one growing season.
Biennials germinate the first
season, then flower, produce
seed, and die in the second
season. Perennials have an
indefinite life span. They are
generally slower to develop
than annuals, but flower and
set seed year after year.

Trees, shrubs, and vines are
also perennials, but their soft,
young stems quickly become
tough and woody, forming
trunks covered with protective
bark. These woody perennials
are either deciduous or evergreen. Deciduous plants cope
with environmental stress (such
as cold winters or extreme
heat) by losing their leaves, a
new set appearing when
conditions improve. Evergreen
trees and shrubs have tough,
needle-like, or leathery, leaves
that remain on the plants
through stressful periods.

PLANT NAMES

Scientific names assist in identifying those plants that have more than one common name.

THE DISTINCTIVE LEAF *shape and fruit (acorn) of* Quercus macrocarpa *(left) identify the plant as belonging to the genus* Quercus *(oaks).*

Most of us are familiar with the common names of plants. Unfortunately, these can present us with a few problems. Some names refer to quite different plants: in Britain, for example, *Hypericum calycinum*, a low groundcover, is known as rose-of-Sharon, while in the United States, the same common name applies to *Hibiscus syriacus*, a large shrub. Other plants have numerous common names, and, just to add to the confusion, still others have no common name at all.

SCIENTIFIC NAMES

Clearly, there was a need for an orderly way to name plants. To address this requirement, Carl Linnaeus, the eighteenth-century Swedish botanist, developed a scientific classification system. His system gives each plant a botanical name of two words (a "binomial"), in Latin or Greek. The first word is the plant's genus, which associates it with a group of similar plants; the second, its species name, identifies the plant more precisely. For example, oaks form the genus *Quercus*, with over 400 species that vary in size and form, but share the characteristic of fruits called acorns. When we are referring to a specific tree within the genus, we add its species name, as in *Quercus alba*, the white oak.

Plant families are larger groupings of genera (the plural of genus) that have similar characteristics of flower and fruit. The oaks are grouped with the beeches (*Fagus*) and chestnuts (*Castanea*) into the Fagaceae, the beech or oak family. There are six genera in this family, and more than 600 species, but they all have more-or-less pendant, catkin-like flowers that are wind-pollinated, and fruits that are nut-like and enclosed in a tough shell.

A plant's botanical name generally tells us something about it, such as the person who discovered it, where it grows naturally, what it looks like, and so on. Many plants

THE FAMILY TREE *for the Bignonia family illustrates the relationships of genera, species, variety, and cultivars.*

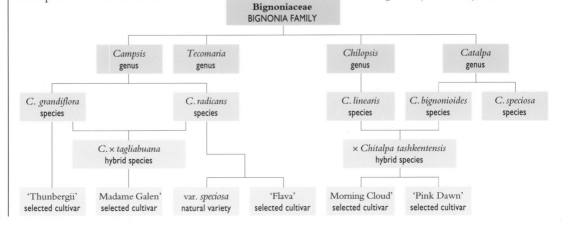

Campsis × tagliabuana. This species displays characteristics of both parents.

Many of our most common garden plants are cultivars (*cultivated varieties*) that have been selected, developed, and propagated by nurseries for their outstanding appearance, size, or vigor. Some cultivars have been found in the wild, while others may be the result of selecting the very best from a group of seedlings in a nursery. They are noted by the use of single quotation marks. *Campsis × tagliabuana* 'Madame Galen', is a selected form of that hybrid species.

THE NATURAL VARIETY Viola *papilionacea* var. *priceana is now known as* Viola sororia *(above). California poppy (*Eschsholzia californica*) (above left) takes its common name from its native region. Common names of plants can be delightfully inventive, such as the romantically named love-in-a-mist (*Nigella damascena*) (below).*

occurring variations from the typical character of a species. *Salvia azurea* var. *grandiflora* is a blue sage with larger flowers than normal.

Hybrid plants are the result of natural or artificial crosses between different species (and occasionally genera) and are identified by a multiplication sign. For example, the trumpet creeper (*Campsis radicans*) of North America has been crossed with an Asian cousin (*Campsis grandiflora*) to produce the hybrid species known as

carry the names of famous botanists. Linnaeus named the genus *Magnolia* after the seventeenth-century French botanist, Pierre Magnol, whose idea it was to group related species into a single genus.

Some names refer to the geographical or ecological origin of the plant, as in *Aesculus californica*, a buckeye from California, or *Nyssa sylvatica*, the sour gum of the eastern forests (*sylvatica* means "of the forest"). Other names describe the physical characteristics of a plant. *Salvia azurea* (blue sage) has blue flowers, *Viguiera multiflora* (showy goldeneye) has many flowers, and *Parthenocissus quinquefolia* (Virginia creeper) has leaves divided into five leaflets. *Carnegiea gigantea* (saguaro) is a large cactus.

ADDITIONAL DIVISIONS
Subspecific names (subspecies, variety, and form) are on occasion given by botanists to plants that display naturally

NATIVES *versus* EXOTICS

Every wild plant is native somewhere; identifying the plants that are native where you are is the important thing.

North America's extremely varied topography and climate have given rise to a rich and diverse plant population, providing a most wonderful range of plants to be considered for the natural garden. Each native plant is an integral part of a community of interdependent plants and animals called an ecosystem. Some of these natives have evolved as generalists. Tolerant of a wide range of conditions, they are to be found naturally over broad sections of the entire continent. Others are specialists, thriving only under very specific conditions of climate and soil, and often limited to a very narrow natural range. It is important to understand the natural distribution of a native plant to determine its suitability for your garden situation.

NATIVE HERE

A native plant is one that has grown naturally in a particular area without human influence or interference. There are degrees of "nativeness", however. A plant such as a rush (*Juncus* spp.) might be native to a continent (North America), to a geographic region within the continent (the West Coast), to a political boundary within the region (California), to an ecosystem (Central Valley grassland), to a particular habitat within the ecosystem (streamside community), or to a particular site within the habitat (a single spring feeding the stream). The fact that a plant is native to California, gives us few clues to its cultural needs; identifying its specific natural habitat will reveal the local conditions of soil and climate that foster the successful growth of the plant.

INVASIVE, INTRODUCED PLANTS, *such as purple loosestrife* (Lythrum salicaria) *(below left) and ox-eye daisy* (Chrysanthemum leucanthemum) *(below right), make attractive plantings in the natural garden, but only if their spread is controlled.*

NATIVES AND EXOTICS *make delightful dried flower arrangements (above). One of the best butterfly-attracting plants is the exotic butterfly bush (Buddleia davidii) (right). Dandelions (Taraxacum spp.) (below right) are typically considered a weed, although they are a useful nectar source for bees and butterflies.*

NATIVE THERE, INTRODUCED HERE

European and Asian peoples migrating to North America brought plants they were accustomed to growing in their homelands. These introduced plants, called "exotics", often found the conditions to their liking and grew beautifully, eventually becoming some of our most familiar—and valuable—garden plants. Some of the more adaptable species escaped from cultivation, invaded local natural areas, and in effect turned into "naturalized" resident species. Plants native to one part of the United States may also be considered exotics in another part: though common in California as a street tree, the tulip tree (*Liriodendron tulipifera*) is actually a native of the eastern woodlands.

PROS AND CONS

On the whole, it is often easier to grow plants that are native to your area than introduced species because they will be adapted to local climate and soil conditions; most native species will be accustomed to the natural rainfall, so will need little or no supplemental irrigation *once established* in your garden. They are also likely to be resistant to pests and diseases within their natural habitats, reducing the need for pesticides or other control measures. However, a rare native plant may be so because it has very special needs (such as specific mineral content in the soil), which may be difficult to satisfy in a garden. Plants native to a particular region will not necessarily be adapted to all garden settings within that region.

Many introduced plants can be quite adaptable, and may be just as suited to your garden as the local natives. To avert the danger of exotics escaping from your garden, avoid those that produce copious quantities of wind-dispersed seeds, particularly if you garden within visual range of undeveloped natural areas. Check with your local native plant society for their list of escaped exotics that should not be planted in your area.

The majority of plants in The Backyard Habitat (pp. 100–272) are native to the region in which they are placed; a few are exotics that have been found to behave themselves and not escape into the surrounding country-side. Each of the plants has been chosen for its value in creating wildlife habitat, as well as for its adaptability to cultivation in a variety of circumstances; many of the natives listed will, in fact, thrive in gardens well outside their natural range.

WHAT IS A WEED?

Nearly any plant may be considered a weed if it is growing where it is not wanted. Even some of our native plants become weeds by invading farmers' fields or by spoiling the look of a groundcover bed. The value of all weeds is that they cover the ground quickly, reducing the danger of soil erosion. Many also play an important role in providing nectar for butterflies and other insects, and foliage for their larvae.

WEATHER WISDOM

The wise gardener has a good idea of what to expect from the climate and plans a garden that will make the most of prevailing conditions.

All aspects of the climate influence the growth of plants: temperature, precipitation, humidity, wind, and sunlight. Native plants are found growing naturally where soil and climatic conditions are most suitable for them; those that are naturally widespread are likely to be more tolerant of a range of conditions. Plants that are introduced have preferences for climate and soil, as well, but will often adapt readily to a variety of conditions. To understand how best to use both native and exotic plants, you must study their needs and determine if your conditions are suitable for their good growth.

EXTREMES OF WEATHER *such as frequent high rainfall and storm activity (below) must be taken into account when designing your natural garden.*

HARDINESS ZONES

A plant's ability to endure harsh conditions is known as its hardiness; although soil, water, sunlight, and humidity may affect a plant's hardiness, the most important factor—and the most quantifiable—is temperature. The gardener will therefore want to know what extremes of temperature a given plant can be expected to survive. To this end, the United States Department of Agriculture (USDA) has prepared a Plant Hardiness Zone Map that divides the United States and Canada into 11 zones, the lower numbers representing the colder zones based on *average minimum winter temperatures*. Plant references, including those in The Backyard Habitat, beginning on p. 100, usually provide the zone numbers for each plant to indicate the areas in which it can reasonably be expected to survive.

THE USDA PLANT HARDINESS ZONE MAP *(right) divides the USA and Canada into 11 zones, from zone 1 (below -50°F) to zone 11 (above 40°F). Hardy evergreen trees and shrubs (far right) are suited to areas of extreme cold.*

The USDA map is a good starting point for selecting plants for your garden, but average winter temperatures give only part of the story. Often the timing of cold weather is the constraining factor; an early freeze in fall may kill plants that are still settling in for the cold winter even though they could survive much colder temperatures in midwinter. Likewise, a spring freeze can seriously harm plants which have already started their seasonal growth; native plants are often better programmed to wait until after the last chance of frost to begin their season's growth.

The plants that thrive in winter low temperatures are very much dependent upon cool winter weather to give them a dormancy or rest. In the mildest regions of the country, the same plants may suffer from the lack of winter chill. Conversely, those plants accustomed to a cool summer, will suffer if the summer temperatures are too high.

The USDA map is most effective in the eastern two-thirds of the continent, where there are few mountains to influence the movement of

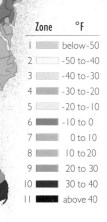

Zone	°F
1	below -50
2	-50 to -40
3	-40 to -30
4	-30 to -20
5	-20 to -10
6	-10 to 0
7	0 to 10
8	10 to 20
9	20 to 30
10	30 to 40
11	above 40

cold air masses. In the western regions, the north–south mountain ranges have a profound impact on the climate, with the proximity to the moderating influence of the ocean among the most important factors.

WESTERN CLIMATE ZONES

Recognizing that winter low temperatures are only one factor in considering a plant's climatic suitability, the editors of the *Sunset Western Garden*

Book have mapped out a much more detailed set of climate zones for the West. This system divides the region from east of the Rocky Mountains to the Pacific coast into 24 climate zones using six key factors: *latitude* (distance from the equator affects sun angle, day length, and season length); *altitude* (elevation above sea level affects intensity of the sun, night temperatures, and season length); *proximity to the Pacific Ocean* (affects humidity,

temperature, rain and wind patterns); *influence of the continental air mass* (affects rainfall and seasonal temperature extremes); *mountains and hills* (influences the movement of oceanic and continental air masses); and the *local terrain* (mostly affects winter temperatures). Indicator plants have been used to define what is likely to succeed in each of these zones. Similar maps are being researched for other regions of the country, but are not yet ready for publication.

MICROCLIMATES

Microclimates broaden the possibilities for gardening beyond that of the general climate. Natural factors such as topography and slope orientation will affect the degree to which the sun heats up a space. North-facing slopes will always remain cooler than those that face south; valley bottoms will collect the heavier cold air at night and be frostier than the slopes just above the valley floor. Mountains have a profound impact on the amount of rainfall a region records; the rising and cooling of moisture-filled air creates high rainfall areas on one side, while a drier "rain shadow" usually develops on the lee side of the mountain.

Natural bodies of water—even quite small ones—will modify the temperatures in the surrounding area. Islands

GENERAL CLIMATE CONDITIONS, *as well as specific variations or microclimates (below), are important when choosing both native and exotic plants for your garden.*

and peninsulas are noted for their moderate temperatures: nearly surrounded by water, San Francisco is generally warmer in winter and cooler in summer than nearby mainland areas. In the Detroit River, Grosse Isle enjoys a climate that is one to two zones warmer than the rest of southeastern Michigan.

Natural vegetation cover also has an influence on microclimates. Forested areas are likely to be cooler and more humid during the summer than nearby grassland areas; in winter they will be warmer, since wind speeds will be reduced. The wind itself can be both beneficial and harmful in a garden; fresh breezes are

important for avoiding foliar diseases, yet excessive and persistent winds tend to destroy branches and desiccate plants. Rainfall will be more gentle in the forest than in the open, and in winter the snow will fall more evenly and last longer. Furthermore, a good cover of snow will act as an insulating blanket, protecting those plants that might otherwise suffer

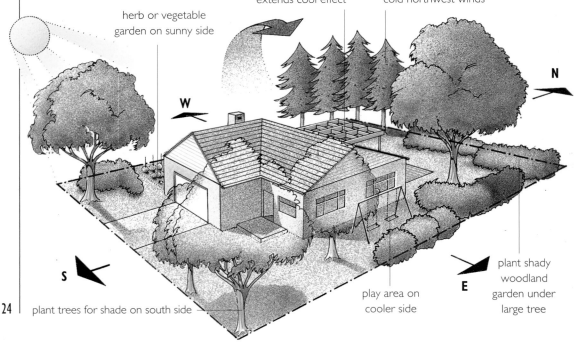

herb or vegetable garden on sunny side

overhead trellis extends cool effect

evergreen trees to block cold northwest winds

W

N

S

E

plant trees for shade on south side

play area on cooler side

plant shady woodland garden under large tree

CHOOSE THE RIGHT PLANTS
for your location, plant for seasonality, and your garden will have much to offer the observer and the wildlife visitor through all the seasons, from summer (left) to winter (below).

from being exposed to extreme cold temperatures.

Human-made features in the garden can cause similar microclimatic effects. Shadows cast by a north-facing wall result in a cool, "north-woods" effect, while sunny, south- and west-facing walls may create near desert conditions on hot days. East-facing walls protect plants from the heat of the afternoon sun, while letting them bask in the morning light.

Ponds and swimming pools will hold enough heat to modify slightly the air temperature around them, though the effects will be insignificant in cold eastern winters. Water that is constantly moving—a natural stream or a recirculating fountain—will have a greater impact during the winter months, and can provide vital moisture for wildlife.

Trees can be planted to enhance or modify the microclimates, by shading areas that would otherwise be exposed to too much sun. Deciduous trees will, of course, let the sunlight through during the winter when they are without leaves, while still providing cooling shade during the leafy summer months. Hedges and shelterbelts of tall trees are often used to control the drying and destructive winds of the prairies and coastal regions; they may also effect a rise in the apparent air temperature. Deciduous trees and shrubs, acting like snow fencing, will help build up the snowpack to insulate the garden for the winter. Evergreen trees and shrubs will provide necessary shelter for many forms of wildlife during the worst periods of winter weather.

RAINFALL
It is as important to understand the local pattern and/or season of rainfall as to know the amount of rain likely to fall on the garden. Most plants grow during the warmer months, and prefer their moisture then; if it doesn't rain, supplemental irrigation may be necessary. In the Mediterranean climate of the far West, most of the rainfall occurs from fall through early spring; native plants in this region are well-adapted to the annual summer drought, and do much of their growing during the cooler, wetter months. The great advantage of depending heavily upon native plants is their natural adaptation to the rainfall of the region; introduced plants should be selected on the basis of matching their water needs with the rainfall in the area.

RAINFALL VERSUS DROUGHT

With the availability of water becoming an increasing issue in many parts of North America, it is important to understand the natural rainfall patterns and to plant accordingly. Droughts are headline news all across the continent, but the definition of drought varies according to the region. A week without rain in Georgia might be considered a drought, while an entire summer without rain is normal for most of California. In the Rockies, several years of below average rainfall are usually necessary before a drought is declared. Clearly, understanding the rainfall, and selecting plants accordingly, is an important approach to successful gardening.

SOME PLANTS *are adapted for dry conditions. The pinyon pine (Pinus edulis) (above) has specialized needles and roots to help it retain moisture.*

THE PROCESS *of* SITE-PLANNING

Planning and design are the cornerstones of a successful garden, whatever its size.

The process of planning a new garden, or remodeling an older one, can be both challenging and fun, but it can also be complex and overwhelming. Stocking up on plants before you are clear about which ones will thrive on your land, and where they should be placed to achieve the desired effect, can be a major mistake. It is important to first become familiar with the conditions of your garden: sun, shade, damp areas, windy spots, and so on.

If you have the time, it is best to take a year to study a full seasonal cycle in the garden (if there already is one) before taking any major steps. This will then give you plenty of opportunities to study the region's wildlife and climate, as well as to identify any key microclimates. If the property

is nothing but bare ground to start, a little sod and some mulch will work wonders in keeping dust and mud under control. (A deep mulch will improve the soil, making it perfect for planting when the time comes.) Go slowly: much of the pleasure of gardening lies in the creative process.

LOCAL SURROUNDINGS

The first step in designing a garden is to look around your neighborhood. Find out what natural habitats still exist, what native plants make up the plant communities, and what wildlife species are common. Make note of well-designed nearby gardens. Find out what plants have been successful under conditions similar to yours. Take photos of the gardens and plants that

The only limit to your garden is at the boundaries of your imagination.

THOMAS D. CHURCH (1902-78), American landscape architect

you like best. Take time to discuss your view of natural gardening with neighbors to avoid any future conflicts. Inquire at the local planning department about rules and regulations concerning the design of gardens and plant choices in your community.

SITE INVENTORY AND ANALYSIS

Even if you are planning to use a professional designer, it will be valuable to work through the site-planning

PLANT FOR SEASONALITY *(right) to make the most of your site. A water source (above) is an important feature of any garden, even a small garden for people living in the midddle of a large*

city, such as New York (top).

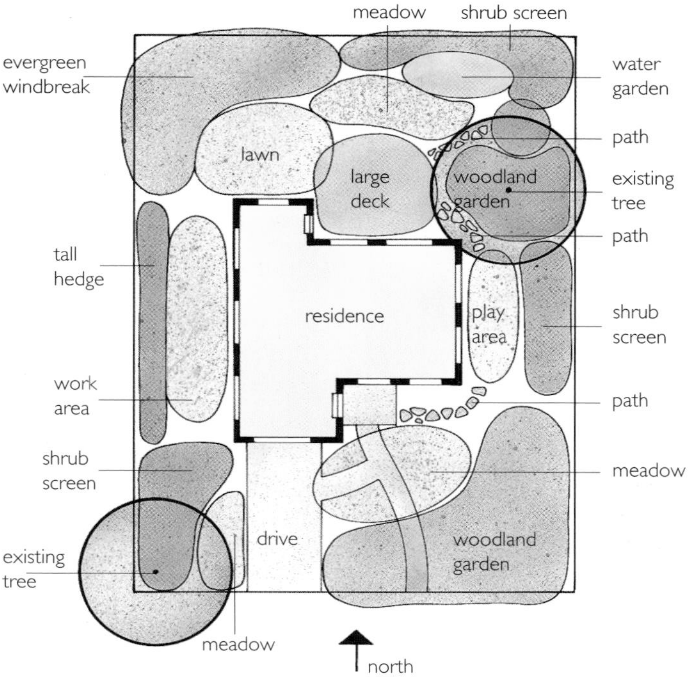
meadow
shrub screen
evergreen windbreak
water garden
path
lawn
large deck
woodland garden
existing tree
path
tall hedge
residence
play area
shrub screen
work area
path
shrub screen
meadow
drive
woodland garden
existing tree
meadow
north

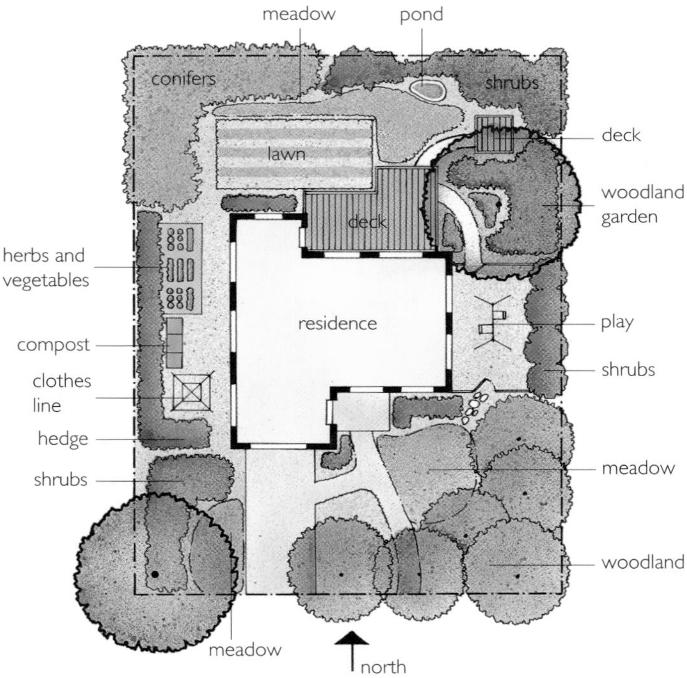
meadow
pond
conifers
shrubs
deck
lawn
woodland garden
deck
herbs and vegetables
residence
compost
play
shrubs
clothes line
hedge
meadow
shrubs
woodland
meadow
north

PERMANENT FEATURES

- structures, including walls, fences, decks, buildings
- pavement, such as paths, terraces, driveways
- utility easements
- views out or in that need blocking or framing
- established trees, shrubs, or other plants
- ponds, creeks, swampy areas, drainage channels
- changes in ground level
- soil types
- resident animals, both domestic and wild

SEASONAL ELEMENTS

- prevailing winds
- sun and shade patterns
- areas where frost collects, or snow lingers
- maximum and minimum temperatures
- rainfall and drainage patterns
- visiting animals

DRAWING PLANS *showing each garden feature or activity is vital in planning your natural garden. A "bubble" plan (top) is developed into a concept plan (left).*

process yourself first. Begin by taking an inventory of your site and analyzing what's there. To do this, you will first need to make a scale drawing of the property, for which you will require a measuring tape, ruler, pencil, graph paper and tracing paper, and a firm board to tape the drawing to. Once your base drawing has been done, cover it with a sheet of tracing paper and draw in or note both the permanent and seasonal features of the site (see box).

Activities and Themes

Take time to make a wish list of the various activity spaces required by you and other household members, such as vegetable or cut flower gardens, a dining terrace, a swing set, or a swimming pool.

You may wish to develop a theme or themes for your garden, such as a garden of a single color scheme. Your choice of theme(s) will affect the layout of your garden, the selection of plants, their placement, and the level of maintenance needed.

Garden Plans

Once you have determined how much area to allow for each activity or feature in the garden, begin exploring ways of placing these areas on the

plan of your property. Do this on tracing paper placed over the original base plan. Include all the activities you've agreed upon, and draw them in as "bubbles" at approximately the size they need to be; don't forget essentials such as trash cans, garden storage, a clothesline, and a compost pile. Consider circulation of people through the garden, identifying areas of greatest impact for steps and pathways. Determine the height of screening that will be necessary to block views or maintain privacy. Do at least two bubble plans on tracing paper before settling on the best one, then develop that one into a conceptual plan which should show in a rough form all the features you want to include in your new garden.

DEVELOPING THE PLAN

With your conceptual plan completed, the next step is to develop the details of the plan, including both hardscape (constructed) elements and plantings. You may wish to consult with a professional landscape architect for the detailed design of the hardscape, or you may prefer to tackle those projects yourself. There are many books that will help you design and build most of the elements you are likely to want.

Take time to prioritize the development of your garden. Although planting may be the fun part for most people, many of the hardscape elements ideally should be installed well before any of the planting commences. Undertake any construction that requires lots

A GARDEN SHOULD PROVIDE
a quiet, natural place of rest for humans and wildlife (right).

Broken slabs of concrete serve as a very acceptable replacement for expensive natural stone in both retaining walls and paved areas. Consider pervious pavement of concrete, crushed stone, or natural stone pavers; you may even wish to install an old-fashioned rain barrel to collect the runoff from the roof of your home.

of work space or access by bulky mechanical equipment. Lay out and install any main irrigation lines that eventually will be connected to sprinklers, drip lines, or faucets. Consider how you will provide water in the garden, and make sure you have a system that is easy to maintain and makes this essential chore as effortless as practical.

CONSIDER THE ENVIRONMENT

In keeping with the theme of a natural garden, pay careful attention to how your garden can be constructed so as not to impact unnecessarily upon the environment. Avoid the use of toxins wherever possible; (see pp. 40–1 for information on nontoxic pest controls). Use construction materials that are free of toxins as well. Consider lumber that is naturally resistant to decay (redwood or black locust), or that is treated with nontoxic preservatives; check with your local organic gardening center for specific products. If possible, use recycled timbers from the local salvage shop to reduce the need to destroy more of the world's forests; do not use old railroad ties if they smell of creosote as this oily preservative is toxic to all living things.

DON'T FORGET THE SIMPLE COMFORTS

A garden is a place where people can escape from the frantic pace of life and be in touch with the natural world. Remember that your garden should serve you as much as it does the wildlife you hope to attract to it. Make certain that there is a comfortable spot to sit, watch, listen to, and enjoy the natural world—the wind in the trees, a chipmunk gathering nuts, or the brilliant colors in a bed of flowers. Bring some of the plantings close to windows you frequently look out of, so that you can enjoy the wildlife from indoors. Search out sculpture for the garden that is visually pleasing; some might even serve wildlife by having one platform for feeding and another for drinking.

GARDEN AREAS *of greatest use can still be attractive (top): here a bark mulch path with railroad ties leads down a shady slope. A John Singer Sargeant painting (left) of American landscape architect Frederick Law Olmsted (1822–1903)*

PROFESSIONAL HELP

Creating a garden is one of the most satisfying of pursuits, yet the complexities of the project can appear overwhelming. Don't hesitate to consult with the professionals who understand the design and development of a garden. Landscape architects are trained in all aspects of landscape design and construction, and can even supervise the installation of your design. Garden designers may lack the academic training of the landscape architect, but often have a greater feel for the natural world of the garden and the plants that are possible within a given region. Check with your local nursery staff for recommendations on plants well suited to your situation. Inquire at local botanical gardens and nature preserves for lists of plants and animals native to your area.

A Natural Approach *to* Planting Design

For visual appeal and practical purposes, it is important to choose the right plant for the right spot.

The final step in the design process is to prepare a planting plan. This will provide the basis for selecting plants for specific purposes: plants to block out unsightly areas or to frame views, trees to provide shade, groundcover to hold a slope, plants for wet spots or hot, dry locations. Keep in mind the size of your garden and do not include large plants unless you have plenty of space. Again, use tracing paper taped over the base plan and the final conceptual plan. Draw a simple circle for each plant, using a diameter equal to the approximate size the plant should be at maturity. Then note one or more species that might work for each circle you've drawn, ultimately choosing the best one for each situation.

Each plant on your plan should have a purpose, and that purpose will dictate how large it should be and what type (annual, perennial, shrub, vine, or tree). The soil conditions and microclimate will dictate the cultural needs of the plant. The plants that are most suitable for screening unwanted views are evergreen shrubs, possibly small or low branched trees, but probably not perennials; or, if the planting space is narrow, plant vines on a trellis. If the site is poorly drained and is in shade, a woodland shrub tolerant of damp soils will be more appropriate than a sun-lover.

PLANTING FOR A NATURAL LOOK

There are few straight lines in nature, and plants seldom grow at regular intervals. A natural effect in the garden, however, is not achieved by planting at random, but rather by following certain design principles that reflect the way plants occur naturally. Be sure to keep these points in mind when arranging your plants:

PLANT EVERGREEN *trees and shrubs (far left) tall enough to screen fences and unwanted views. Planting a variety of trees, shrubs, and herbaceous flowers will enhance your garden (above).*

- group the plants (whether they be flowers, shrubs, trees, or even pot-plants)
- group the groups
- vary the sizes of the plants
- vary the spacing between the plants

When planting in small groups, use odd numbers of plants for a more natural and dynamic effect. This will be particularly effective with larger plants in the landscape. For example, three hemlocks of the same size, when evenly spaced, will look dull and decidedly unnatural; if they are of different sizes, and irregularly spaced, they will

GARDEN SCULPTURE *provides visual interest and a place for birds to rest.*

TO REDUCE MAINTENANCE, *group plants according to their moisture requirements (above and below). Bell your cat to protect birds.*

have the charm of a grove of hemlocks in the wild. Position groups of plants near boulders or tree stumps, at a bend in the path, or at a corner. Follow these examples to achieve the most natural and pleasing effects in your wildlife garden.

SUCCESSION IN THE GARDEN

Nature often takes decades and even centuries to reach a final stage in the development of a natural plant community. This process, known as succession, usually involves a gradual transition from one type of plant community to another. Beginning with bare soil, pioneer species of plants, usually annuals, will quickly sprout from seeds blown in from another area. Long-lived perennials are then followed by shrubs, which soon shade

out many of the perennials. Quick-growing trees enter the scene, out-competing the shrubs, but in turn giving way to larger and longer-lived trees in the final successional stage which is called the *climax* community. Insufficient precipitation, excessive winter cold, or very shallow or infertile soils, may lead to the climax stage occurring earlier in the process, as in the grass-lands of the central Prairies which have few trees.

This transitional process can be mimicked in the natural garden when the site

is relatively bare at the beginning; it is particularly appropriate when the ultimate plan calls for a woodland garden. Plant quick-growing groundcovers of annual flowers and grasses to hold the soil, but also introduce some perennials and shrubs to fill spaces between the sapling trees that will eventually grow into a small forest.

USE LESS WATER

One of the most efficient ways to reduce water use in the garden is to group plants according to their water needs. A large percentage of the drinking water in the United States is used for landscaping; given the cost of water, and the restrictions placed by a growing population, water conservation has become a subject of concern throughout much of the country. Water-efficient landscaping has been shown to cut domestic water usage by as much as half. The Backyard Habitat, beginning on p. 100, identifies the water needs of each plant to facilitate grouping them appropriately in the natural garden.

FINDING *the* BEST PLANTS

*One of the great pleasures in creating a garden
is to search for the plants that will give it character.*

Avoid the temptation to drop by your local nursery or garden center and choose a carload of plants in full bloom to create an instant flower garden at home. Most plants adapt readily to their new site if they are planted as young specimens that have spent a minimum amount of time in a nursery. You should buy just the plants you have identified on the planting plan. Some of our best native plants are seldom to be found in the local retail nurseries, so try to make searching for the right plant an enjoyable part of the process of creating a natural garden. Native plant societies often have plant sales offering excellent material at good prices—frequently, plants not available from other sources.

WHERE TO FIND THE PLANTS YOU NEED

By far the simplest and most expedient way to plant a garden of any kind is to buy young plants at a local nursery. Plants are typically available either bare root, balled-and-burlapped, or in containers; their availability varies with the seasons. Deciduous trees and shrubs are often sold without any soil around their roots (bare root) during the late winter, just as their period of dormancy is about to end. These must be planted at once, or kept in a very cool place to delay sprouting until they have been planted. Balled-and-burlapped plants have their

DIFFERENT COLORS, *textures, and shapes combine to give a garden its character. The daylily (Hemerocallis falva) is a striking choice (top). Pottery, stonework, and paving complement a colorful natural garden (above and left).*

roots growing in a ball of soil wrapped in a woven fabric (not always burlap, however); these are also best planted in early spring before the sun has a chance to dry out the root ball. Today, most retail nurseries sell plants grown in containers, from small plastic pots to large wooden boxes. Container-grown plants are usually available throughout the growing season, and can be planted at any time. Bear in mind, though, that spring and fall are the best seasons.

Most retail nurseries sell only plants that are familiar and popular, the tried-and-true ones that are easy to propagate in great quantities. Less familiar native plants are likely to be available only from smaller specialty nurseries that offer uncommon specimens or plants for the serious collector. Talk to other gardeners about specialty nurseries in your area;

with a little searching, you will discover some surprising sources, and you should receive some excellent advice on the development of your garden. Refer to Plant Sources and Organizations, from p. 276, for a number of specialty nurseries to contact.

Shopping by mail will give you access to many specialty nurseries, some of which sell only by mail. Plants offered will generally be small. The greatest selection will be among perennials and bulbs; most will be sent bare root, some in small containers. With mail-order, reliability does vary so buy only a few plants to start with. Check the list under Organizations for mail-order nurseries located in your region.

Other sources of plants include botanical gardens and arboretums, which now stage regular plant sales to generate revenue for their programs. They usually

TREE PLANTINGS *(right) are often sponsored by local, city, and state agencies. Visit botanical garden shops for plants that may suit your garden (left).*

sell plants propagated from their collections; search out those gardens that specialize in native plants so you can take advantage of their sales.

NATIVE PLANT SOCIETIES

In nearly every state and province there is a native plant society, devoted to the study, preservation, and cultivation of the state's native plants. These organizations serve as excellent resources for learning more about the plants that are appropriate for your natural garden. Many raise funds for their activities through the sale of plants propagated by their members. This is very likely to be the best source of the more uncommon native species. Their meetings and newsletters act as clearinghouses, providing advice on how and where to properly collect seed, and listing the reputable commercial sources. Sharing information with other members can be rewarding.

Some state agencies, such as the forestry department, offer young native plants to the public for revegetating larger properties. Usually, it is only trees that will be available, but water plants, grasses, and wildflowers are occasionally available. The plants are generally small, bare root, and available for a brief time in late winter; the price is usually extremely low, but the selection will be very limited.

COLLECTING FROM THE WILD

Strict regulations govern the collection of plants from the wild, to guard against further pressure being placed on the many native plants that are threatened with extinction. Never dig up plants anywhere in the wild, unless they are growing somewhere where they will otherwise be destroyed, such as a highway project or building site. Even in these situations, obtain the permission of the property owner before digging any plants.

The same regulations apply to nurseries selling native plants. Most now sell native species that have been propagated from cultivated plants, not dug from the wild. Ask where the plants originated; if they are wild-collected, shop elsewhere.

NATIVE PLANT SOCIETIES *are a great source of species native to your region. The proceeds from this plant sale (right), organized by a wildflower association, will fund propagation and conservation of wild plants.*

GROWING YOUR OWN WILD PLANTS

Creating a new plant from cuttings or root division is a rewarding part of creating a natural garden.

Though it is decidedly simpler to plant a garden with young plants you have purchased from a nursery, the reality is that the natural gardener may have to search for cuttings or seeds of some of the desired native plants. In many cases, that search will uncover the plants in a neighbor's garden, the local nature preserve, or a mail order seed catalog.

NEW PLANTS FROM OLD

With a little research and a bit of practice, you will be able to propagate new plants from those in your friends' gardens, at virtually no expense. You may also decide to propagate the few plants you have purchased from a nursery, to increase their numbers in your own garden. There is a variety of ways of propagating plants:

rooting pieces of stems, roots, or leaves; layering stems; dividing clumps; and separating young plants ("pups" or offsets) from a parent plant. Timing is of critical importance with cuttings, layering, and dividing; pups can usually be separated at any season. Ask the staff of the local botanical garden, or the volunteers for the native plant society, for the best technique for any species you wish to propagate (see also the general gardening references in Further Reading, pp. 274–5).

GROWING FROM SEED

Growing your own wild plants from seed is an ideal way to learn about the complexities of local flora, and in some cases may be the only way to obtain particular plants for your natural garden. Fortunately, an increasing

number of seed companies now offer seed of local plants. Limited seed collecting in the wild is permissible, providing you have the landowner's consent; the generally accepted policy is to collect no more than 10 percent of the seed available, leaving the rest to maintain a steady population in the wild.

Make certain that the seed you are about to collect is ready to be harvested. Many seeds become dry and hard inside pods or capsules that have turned crisp and papery as they are about to release their ripe contents. Other seeds reach maturity inside soft, fleshy fruits that are intended for eating before the seeds are actually released into the environment.

To germinate seeds, it helps if you understand the

DIVIDING CLUMPS OF PLANTS *is a useful method of propagation (above). Growing your own plants from seed (right) is the least expensive way to obtain a large number of plants.*

A COLDFRAME *(far left) is a mini, unheated greenhouse. Grow seedlings and cuttings in small containers (left) until they are ready for planting out. Many native plants make colorful cut flower arrangements (above).*

natural process that breaks a seed's dormancy and to mimic that process before the seeds are sown. In the wild, seeds are dispersed from their mother plant at a carefully programmed time to ensure their best chances of germination. Seeds need moisture, warmth, a growing medium (soil), and sometimes light, to germinate and begin growth as a new plant.

Some seeds sprout almost immediately after coming to rest on moist soil. Others will lie dormant for a while. This dormancy is broken naturally by a variety of environmental conditions; for example, cold winter temperatures work on species that are native to northern regions with cold winters. (The process of simulating winter by giving seeds a period of cold, damp

conditions is referred to as stratification.) Streamside plants often need to be soaked for long periods to break dormancy (see also Further Reading, pp. 274–5).

Seeds of most annual and many perennial wildflowers can be sown directly in the garden. Some perennials and bulbs, and most woody plants, are best started in small containers. Germination requires a location with plenty of light. A sunny window-sill indoors may suffice, or a fluorescent light set-up, but a coldframe or greenhouse will provide the best conditions. Seedlings will need to be separated into individual pots and grown on for a period of time until they are large enough, and the weather conducive enough, for them to be planted in the garden.

HEIRLOOM PLANTS

In the push to breed bigger and better flowers and vegetables, nursery staff and gardeners have often forgotten about perfectly delightful and potentially valuable forms of the same species, a few of which are in fact our native wildflowers. We call these older forms *heirloom* varieties, and there is now a flourishing movement around the world to maintain these forms in cultivation. In general, this requires the collecting of seed from plants in gardens and continually dispersing the seed to new growers; those plants that are best adapted to the specific conditions in which they are growing will be the healthiest and most productive of seed.

While relatively few of these heirloom varieties actually fit into our natural garden, maintaining a genetic diversity is an important means of ensuring the strength and adaptability of any group of plants. The most healthy natural garden depends not upon genetically identical clones, but on seedling-grown plants of each species.

SOILS, *from the* GROUND UP

A well-structured, nutrient-rich soil promotes

the best plant growth in the natural garden.

The great chain of life begins and ends with the soil. Nutrients and water in the soil are the raw materials that give life to various forms of plant life. Plants in turn provide food for other life forms, ranging from bacteria to insects to birds and mammals, including humans. Each of these life forms ultimately returns to the soil upon death, providing food for yet more organisms, both scavengers (beetles and centipedes) and decomposers (earthworms, fungi, and bacteria), which will break down the once-living tissue into its component chemical elements—the essential nutrients that will again give life to new generations of

RICH, ORGANIC SOIL *(top) stands as the important foundation on which the natural food chain (below) is built.*

plants and animals. Fostering a healthy soil environment to benefit all these organisms is a basic goal of the natural garden.

SOIL TYPES

Soils are classified in two ways; the texture refers to the size of the mineral particles derived from decomposed rock, and the structure refers to the aggregation of those particles into "crumbs", held together by organic matter. The pore spaces in between the soil particles are important as they serve to aerate the soil and yet hold moisture in which the various nutrients necessary for plant growth are dissolved.

Fine-textured soils are either clay or silt, both of which have very tiny particles. By virtue of their small size, they pack together tightly, leaving small pore spaces that easily absorb and retain water to the exclusion of air; the flatter clay particles pack very solidly, resulting in very slow

water penetration. Water penetrates coarse-textured sandy soils immediately and drains through very quickly, leaving little behind to hold any dissolved nutrients.

Few soils in nature are of only one texture; most are combinations of particle types, such as sandy clays and silty sands. The best is a combination of sand, silt, and clay,

EARTHWORMS

Earthworms do a great job in improving soil quality. They feed largely on surface organic material, excreting pellets that fertilize the soil (by changing the nutrients into a dissolvable form that plant roots can take up easily), and through their tunneling, aerate the soil, thus enhancing drainage. When there is plenty for them to feed on, they reproduce at a great rate: there can be up to half a ton of worms per acre (0.4 ha) in soil under the best of conditions. Worms are also beneficial in the compost pile as they speed up the decomposing process. Your local organic gardening center will likely have worms available for adding to the compost or the garden.

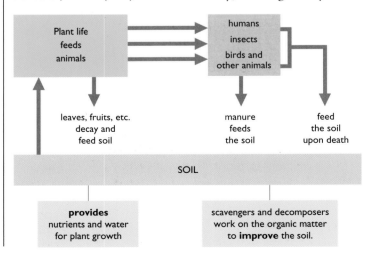

Plant life feeds animals		humans insects birds and other animals	
leaves, fruits, etc. decay and feed soil		manure feeds the soil	feed the soil upon death
SOIL			
provides nutrients and water for plant growth		scavengers and decomposers work on the organic matter to **improve** the soil.	

called loam; its mix of pore sizes permits penetration and drainage of water, but retains some moisture in the soil, providing sufficient nutrients for good plant growth.

The natural activity of decomposing organisms helps cement soil particles together into crumbs that characterize the structure of the soil. These crumbs create a soil with both small pores to hold water and dissolved nutrients and large pores through which air can pass. A healthy soil, with a good structure that is capable of nurturing sturdy plants, may contain as much as 50 percent organic matter.

FALLEN BRANCHES *(below) can be left to decay and form part of the soil-enriching humus. Fungi growing on the branches are attractive to garden wildlife.*

IMPROVING YOUR SOIL

Evaluate the soil: are healthy plants growing in it now or is it barren? Does water pond when it rains or does it soak in quickly? Is it easy to dig or hard as a rock? Simple home soil tests can tell you if the soil is lacking in nutrients, or if it is exceedingly acidic or alkaline. Contact your local state-run agricultural extension office for advice on professional testing of your soil. If the soil has been relatively undisturbed by construction or previous gardening activities, it may need little or no improvement, particularly if you focus on native plants that are adapted to the existing soil.

For most situations, the incorporation of organic amendments will likely make a sufficient change in your soil. A vast array of organic amendments is available in different parts of the country. The best amendments are those that offer a variety of organic particle sizes, with the organic material derived from both herbaceous and woody plants. Their advantage is that they take longer to decompose into their chemical components.

Well-aged animal manures are an excellent source of organic matter, but beware of weed seeds. Sphagnum peat moss is another top quality amendment, but will make

the soil more acidic if used in quantity; mountain peat is a poor substitute and should be avoided as its extraction is destroying fragile mountain wetlands. Used mushroom compost, redwood sawdust, and rice hulls, are locally available amendments of value to the gardener; check with your local garden supplier for such regional materials. Local waste-treatment facilities are often a source of organic amendments that are made from sewage sludge or from composted tree trimmings and garden refuse.

Composting is an excellent way to improve your soil, from a simple home pile of garden trimmings that is easy to gather up on a regular basis to more sophisticated systems for the avid gardener, some of which produce useable compost in just a few weeks.

MAINTAINING *the* GARDEN NATURALLY

Time needs to be set aside on a regular basis to provide for the care of your plants.

Maintaining a garden naturally means understanding the natural processes that keep any community of native plants healthy and growing, and adapting them for your own maintenance program. Make certain that the plants you select are appropriate for the environment in which you will be planting them. Observe their growth patterns and learn what you can expect from each at different seasons. Study how nature provides moisture and essential nutrients, and attends to pest and disease problems. Bear in mind that some form of succession will occur naturally if you do not exert a little energy at least toward maintaining the community you have planted.

A HEALTHY SOIL ENVIRONMENT

Nature maintains a healthy soil environment by recycling old leaves, twigs, flowers, and fruits into a layer of organic mulch, which *reduces* evaporation from the soil, erosion by wind and water, and compaction from foot traffic; *moderates* temperature fluctuations in the soil; *hinders* the sprouting of weed seeds; and *releases* nutrients back into the soil as earthworms and other organisms decompose the organic matter. Mulch also provides habitat for insects and other ground-feeding organisms. Recycle your own garden trimmings into a deep mulch around your plants; use home shredding machines to chop tough, woody branches. Avoid burying the leaves of perennials and annuals with mulch, and keep the mulch

Nature knows how to produce the greatest efforts with the most limited means.

HEINRICH HEINE
(1797–1856), German poet

PRUNING PLANTS *(left) is an essential part of garden maintenance to increase flowering or fruiting. A layer of mulch, made from leaves and newspapers (below), encourages the growth of healthy vegetables and flowers such as* Viola rotundifolia *(right).*

away from direct contact with the trunks of trees and shrubs, particularly in the arid West. In desert regions, a mulch of small pebbles will mimic the coarse sand layer that tops most desert soils.

Many of our native plants do better without excessively fertile soils. If soil tests dictate the addition of fertilizers, provide them in the form of organic products that usually offer a range of trace minerals

beyond that found in product that is chemically manufactured.

CONSERVING WATER

If you have planted your garden with nearby natives, and have grouped them according to their water needs, you will very likely need to worry little about providing supplemental irrigation. Even those plants that require extra water, such as stream-side growers, are programmed to survive periodic reductions in the soil moisture level, and will spring back when the rains come again.

Admittedly, most of us want our plants to thrive, not just survive under our care. An irrigation system may prove necessary to get new plants established, and for an occasional boost of water once established. For many, hose-end sprinklers will be sufficient; for others, a complete underground system may be needed, but avoid total dependence upon an automated system. Take time to observe your plants and their water needs; most plants will begin to show signs of water stress long before they dry out completely, giving you plenty of time to provide the necessary moisture. Look for sprinkler heads that release low-volume streams of relatively large droplets; you will then reduce the loss to evaporation, and will avoid eroding the soil's surface. Irrigate in the early morning

to avoid evaporation by the sun and loss due to winds, and to avoid damp, disease-prone leaves in the evening. Water deeply, but infrequently, even during the establishment period, to encourage deep rooting that will allow plants to survive water shortages easily. Water plants that are widely spaced with a drip, or trickle, system of tiny emitters that slowly release measured quantities of water where it can soak down directly into the root zone of each plant.

PRUNING TO SHAPE AND CONTROL

Prune plants to remove any dead, damaged, disfiguring, weak, diseased, or crowded stems or branches; to increase the flowering or fruiting; to direct the growth or shape of a plant; and, lastly,

WEEDING *(top)* **AND WATERING** *(above) are important in maintaining a healthy garden environment. Most natives, however, require little extra water beyond the natural rainfall.*

to control the size of a plant. Select plants that will stay within the space you have allocated for them on your planting plan. When shaping is called for, avoid shearing into dense, unnatural shapes that block the entry of birds and other wildlife. Always prune to enhance the natural shape of your plants. You can also control the growth of part of your garden to maintain it at a particular successional stage.

A Natural Approach *to* Pest Control

The occasional chewed flower is a minor blemish in a garden planned for wildlife as well as people.

The more diverse the composition of a garden, the more examples of producer–consumer, predator–prey, and parasite–host relationships present themselves. Each demonstrates that nature will take care of itself, a model for the natural gardener who will seldom find the need to resort to chemicals to control occasional outbreaks of pest or disease among the plants.

Integrated Pest Management

Integrated pest management (IPM) is a systematic approach to the control of pests and diseases that is based on the premise that a healthy plant, well-suited to its environment (as native plants usually are), will seldom suffer from problems caused by pests and diseases. When problems arise, IPM dictates that the pest or disease be carefully identified and monitored to determine the seriousness of the situation; when a problem exceeds an acceptable level, a series of nontoxic controls will be tried before resorting to any chemical solutions. These "safe" controls include cultural, mechanical, and biological strategies; if these fail to halt the problem and chemicals become necessary, those derived from natural sources are emphasized.

Start with cultural practices aimed at improving the overall health of your plants, making them less prone to pests or diseases. Watch for signs of drought stress or saturated soil, too much or too little sun, excessive drying winds. Fertilize if a nutrient deficiency is observed. Remove a mulch if it harbors snails, slugs, or other pesky creatures, or contains diseased leaves or petals that may be releasing spores to infect more plants. Eliminate weeds that may also be harboring pests or spreading diseases. Cut off infected leaves and branches and dispose of them to reduce the spread of a disease.

If insects are the problem, consider mechanical controls such as traps, cages, barriers, or other devices that will keep them off your plants. Sticky bands on tree trunks, and bright yellow sticky flypaper attract a variety of insects. Stiff hose sprays will wash many insects away. Hand collecting slow-moving pests will remove many if you are persistent; most pests are feeding at night so you need a flashlight to find them.

An increasing number of biological controls are now available to the natural gardener. Birds and bats, frogs and toads eat many insects, including the pesky mosquito. Beneficial insects, such as ladybugs and praying mantis, can be entertaining as they feed on aphids and other insects. Tiny predatory wasps, mites, and nematodes—virtually invisible to the naked eye—feed on, or parasitize, even tinier creatures. Some may already be active in the garden, but supply houses can provide you with eggs, larvae, or adults by mail. Also available are pathogens such as bacteria, fungi, and viruses that attack and kill the pests that ingest them; others work to control diseases.

When all other measures fail, you may be forced to

CHEMICAL CONTROLS *would kill birds, spiders, and insects that are themselves natural pest controllers, such as insect-eating birds (above).*

resort to pesticides. Look for those that are derived from natural sources; use them sparingly and carefully, targeting the specific problem that concerns you.

Insecticidal soap sprays work well on soft-bodied insects such as aphids and mites. Fungicidal soaps control rusts, mildews, leaf-spots, and other surface diseases. Soaps like these are not toxic to mammals, and biodegrade quickly in the environment; some plants may be harmed by them, however, so it is best to test them first on a single plant.

Spray horticultural oils at any time of the year to kill pest insects in the egg or larval stage. They have a very low toxicity to humans and wild-life and break down quickly.

Botanical pesticides, derived from plants that are known to be toxic to insects, can be used effectively to control specific insects; some, however, are broad-spectrum, and will kill all insects that come into contact with the pesticide. The toxicity of these naturally based products varies, some being fairly toxic to both humans and wildlife, but they generally biodegrade rapidly upon contact with the soil or when exposed to the sun. Always follow the directions carefully when using any type of chemical.

Remember that a few pest insects in the natural garden are unlikely to create a problem, and may even be a source of food for other forms of wildlife you wish to attract to the garden.

WEED CONTROL

The secret to controlling weeds is in not letting them get started in the first place. Keep the soil covered, either with mulch or with other plants that shade out any seedlings attempting to grow. Avoid pre-emergent or post-emergent chemical weedkillers; they may leave toxic residues that could affect other desirable plants. Pull weeds as soon as they are spotted; prevent any weeds from flowering and setting seed. An exception is some weeds which actually can be considered an important addition to the garden, attracting various forms of wildlife by offering nectar, pollen, seeds, or foliage for larvae.

Be sure to eliminate the seedlings of trees or shrubs that may be native to the area but are invading a meadow or prairie garden, if you wish to maintain that grass-dominated community. Likewise, concentrate on reducing the more aggressive annual and perennial wildflowers that are inclined to take over, simply by dead-heading the flowers before the seed is produced.

BENEFICIAL INSECTS, *such as ladybugs, eat pests like aphids (far left). An early-morning spider's web (left).*

41

PLACES *of* INSPIRATION

A small windowbox or a vast natural landscape can provide ideas you can adapt for your garden.

Inspiring visual ideas are all around you. Analyze the "ingredients" that make a garden or vista attractive—the many shades of green in the foliage; the color and shape of the flowers; the relative scale of the plants; the subtleties of texture. These observations will prove invaluable as you select plants native or adapted to your region for your garden.

Utilize your local botanical garden as a valuable source of visual stimulation and information. The settings and groupings of plants featured there will give you a feel for how various plants could work in your situation. Botanic gardens often have plants for sale, and you will have access to informed staff and, in many cases, a library.

A FRONTYARD *becomes a prairie in miniature (top) with purple coneflowers (*Echinacea purpurea*) and black-eyed Susans (*Rudbeckia hirta*). A swathe of lupines surrounds Lake Marion in Grand Teton National Park, Wyoming (above). Simple, yet stunning, California poppies (*Eschscholzia californica*) dominate the landscape in the Mojave Desert (right).*

*To be beautiful
and to be calm;
without mental fear,
is the ideal of Nature.*

RICHARD JEFFERIES (1848–87),
English naturalist and writer

FALL COLORS *(above), add a
welcome warm touch at Mount
Katahdin, in Baxter State Park, Maine.
Woodland waterfalls at Roaring Fork
River, Great Smoky Mountains National
Park (above left). Desert plants (left) at
the entrance to Boyce Thompson
Southwestern Arboretum, Arizona.*

Nature, in her blind search for life, has filled every possible cranny of the Earth with some sort of fantastic creature.

JOSEPH WOOD KRUTCH (1893–1970),
American critic and naturalist

CHAPTER TWO

GARDENING *to*
ATTRACT WILDLIFE

CREATING *the* WILDLIFE GARDEN

Our curiosity about the natural world

and a sense of kinship with other living creatures

create a desire to observe wildlife at close range.

For many gardeners, attracting a variety of animals into the garden is one of the major goals of their efforts. Gaily colored birds and butterflies flitting about the backyard are a popular focus of interest, but the possibilities include less obvious creatures, such as a spider curled up in a flower or a shy lizard making a fleeting visit to the water's edge— such creatures play their own, important roles in a balanced garden community.

Success in drawing wildlife into the garden lies in understanding and providing for its needs: food, water, shelter, and nesting sites. At the same time, integrating habitat with amenities for people—space for relaxation, entertainment, and recreation—enables you to enjoy the sights and sounds of garden wildlife.

ASSESSING YOUR GARDEN'S POTENTIAL

It's useful to begin with a plan, both in the mind's eye and on paper. Initially, it's helpful to view your garden as a unique combination of assets and limitations, reflecting its location, existing site conditions, and potential for improvement. See The Process of Site-Planning on pp. 26–9.

BRIGHTLY COLORED BUTTERFLIES
such as this gulf fritillary are a popular visitor to wildlife gardens. If you want to attract butterflies to the garden, resist the urge to combat problem insects with inorganic pesticides: they are likely to kill the innocent along with the guilty.

Since regional climate varies so dramatically, and since some animals occur only in certain parts of the continent, the geographical location of your garden will determine the wildlife you might realistically hope to attract and which plants are likely to thrive. Refer to The Backyard Habitat, starting on p. 100, for some direction on the particular animals and adaptability of specific plants in your region. More detailed and comprehensive information can be found in other books on North American wildlife and plants (see Further Reading on pp. 274–5).

TARGETING DESIRABLES AND UNDESIRABLES

Start to create a list of the animals you wish to attract, and note their habitat needs and behaviors. You should also consider those animals you might want to discourage, such as raccoons or deer. Be sure to consider the concerns of neighbors, who might resent efforts to support certain animals they consider to be a nuisance.

Avoid sentimentality when you discover that a cherished garden animal is preyed upon by another. Whether it's a snake that eats toads or a bird that eats bees, recognize the valid role that predation plays at all levels in the garden, as in the wild, and take pride in fostering a balanced community of animals and plants.

GARDEN HABITATS *Providing cover near amenities such as ponds (above) will enhance their attractiveness. A garden linked to woodland habitat (right) will attract forest-dwelling species.*

Take a look at which plant species grow best in your neighborhood and in nearby wild habitat to help you decide what to plant to attract your favorite wildlife species.

Assess your garden's current performance in providing habitat essentials. By identifying those plants currently favored by wildlife, you can decide which plants to repeat and which ones deserve to be phased out or relocated.

THE GARDEN AS HAVEN

To survive, animals must contend with predators, the elements, and competition from other animals for food. Wildlife-sensitive design creates patches of animal-friendly habitat—secure space for breeding, resting, drinking, bathing, and feeding.

Plants constitute the most important element in meeting the varied needs of wildlife. Trees, hedges, and shrubs moderate wind, sun, and evaporation to create different garden microclimates. Branches and foliage provide shelter, cover, and nesting space. Plants are the main source of food in the garden, and animals have evolved feeding habits to take advantage of plant pollen, sap, nectar, leaves, flowers, buds, seeds, fruits, nuts, and even the wood fiber itself. In some cases, the plants support small insects that more conspicuous wildlife, such as dragonflies and warblers, require in their diet.

CREATING HABITAT NICHES

Generally, the more wildlife habitats, or niches, you create, the more wildlife species will be attracted. Be sure to establish sufficient habitats to satisfy the basic needs of the animals you want to attract; this may vary from a pool the size of a washbasin for certain small frogs, to acres of critical habitat for mammals.

… a taste for the beautiful

is most cultivated out

of doors, …

HENRY DAVID THOREAU (1817–62), American essayist, poet, naturalist, and philosopher

IN YOUR NEIGHBORHOOD

*The real rewards of natural gardening come from
learning the details of what matters to local wildlife.*

Each region, and location within a region, has plants with value for native wildlife species. Native plants usually are the most attractive plants for wildlife, but they don't always reign supreme. Some introduced plants can be highly popular with local wildlife. For example, the Russian olive (*Elaeagnus angustifolia*), in the Denver, Colorado, area, seems to be a favorite of nearly every raccoon, bird, squirrel, and deer. But such plants may constrain wildlife diversity, or may be invasive in the garden.

CONSIDERING THE SITE

Each site comprises its own unique combination of assets and limitations, and even in the middle of large urban areas, some of the best wildlife dramas occur. An established, self-seeding core of plants is the most welcome asset a garden might have to offer. A grove of trees nearing maturity provides considerable habitat value that even extends to its declining years, when hollow trunks provide nest cavities. Brush piles, naturally decaying leaf litter, and other organic materials enrich soils and create feeding and shelter opportunities for many kinds of different wildlife.

Physical features on the site can be equally important: a natural water feature, such as a pond, stream, or seep, is very valuable, providing water for drinking, and habitat for specialized wetland species of plants and animals. Moderate slopes and varied terrain help to divide the garden into different habitat zones.

Physical extremes can limit a garden's potential. A very small size obviously limits the ability to create habitat variety, but even a well-stocked window box draws a host of visitors. Soils that are very sandy or slow-draining limit the plants that can be grown without a program of soil amendment. (See Soils, from the Ground Up on pp. 36–7 for more details about improving your soil.) Dense soil also directly limits the ability of burrowing animals to create homes, though this defect can be useful in deterring problem animals, such as gophers and woodchucks.

Neighborhood factors can also diminish a garden's

CREATING SUCCESSFUL WILDLIFE HABITAT

- Plant to suit the natural topography: low points are suited to moisture-loving plants or water features; a high, sloping bank is ideal for dry rockeries.
- Shield the garden from winds, especially cold, winter ones.
- Introduce food plants favored by animals you have specifically targeted.
- Strive for diverse plantings, but avoid a hodgepodge of unique or isolated specimens.
- Coordinate habitat with water use: lush, irrigated habitat close to the house; drier farther away.
- Locate some food plants and feeders in the more secluded corners of the property to attract the shyer, ground-dwelling animals.
- Don't use inorganic pesticides: they will unbalance the predator–prey relationships natural gardening helps to foster.
- Arrange plantings of shrubs to lead shy wildlife closer to the house and patio for convenient viewing.
- Supplement food plants with feeding stations; boost nesting opportunities with nestboxes.

EVERGREENS *such as Engelmann's spruce (Picea engelmannii) (top) will provide food in the form of cones for birds and small mammals. The fruits of some plants are highly appealing to many birds. This American robin (left) is feeding on whole cherries (Prunus spp.).*

FLOWERING PLANTS
Increasing the number of flowering plants in the garden (right) will attract many more wildlife species, particularly pollinators, such as this hummingbird moth (below).

potential. Isolation from sources of wildlife immigration is a serious concern, though less so in the case of particularly mobile animals such as birds. A background of heavy urban noise constrains animals dependent on vocal communication, and will deter the more wary roaming mammals, in particular, from entering the site. Berms, walls, and other strategically placed structures can help to alleviate noise pollution.

POLLINATION AND OTHER PARTNERSHIPS

The relationships between local animals and plants are a fascinating part of planning a wildlife garden, and you can organize your planting schemes to encourage such partnerships to occur in the garden. Cooperative relationships, such as hummingbird

pollination, in which all participants benefit, is one kind of partnership, but there are also natural predator–prey relationships in which the system, not the individual, benefits. For example, ladybugs and their larvae will devour the black aphids that appear on plants such as big sagebrush (*Artemisia tridentata*) in spring.

The absence of pesticides allows insect pollination to take place unimpeded. In the case of the more open, bowl-like flowers, a variety of insects cooperates with the flowers in exchange for pollen

CERTAIN PLANTS *will act like magnets to local bees and butterflies. The blue ceanothus 'Julia Phelps' seen growing here (above) will be highly attractive in the West Coast region.*

or nectar. Flowers such as the penstemons are more selective: their corolla is easily penetrated only by the long tongues of hummingbirds. Conversely, roses offer pollen alone, appealing to bees, in particular, and hummingbirds not at all. Pollination partnerships can be amazingly precise, as in the case of certain yucca species that are pollinated solely by the yucca moth.

FOOD, WATER, *and* SHELTER

By deliberately focusing on the various habitat essentials,
you can make the most of any garden's wildlife potential.

Plants, of course, provide the main source of food in the garden, whether in the form of foliage, sap, nectar, pollen, seeds, berries and soft fruit, or nuts. Aim to have food available from plantings and feeders every month of the year: this will attract a diverse range of wildlife species, and birds and mammals will learn where to go when the going gets tough. In severe winter, birds' survival may depend on the day-to-day availability of extra food.

THE BARE ESSENTIALS *Many songbirds, such as purple finches (top), appreciate a source of calcium, which can be provided by crushed eggshells. Probably the best-known North American bird, the American robin is sure to visit a cooling birdbath (below).*

FEED THE BIRDS

Installing a birdfeeder will boost your garden's capacity to support songbirds. Situate it in a visible location, a minimum of 4½ feet (1.4 m) above ground, within 20 feet (6 m) of dense evergreens where the birds can readily find cover. While feeders that provide food on an open platform are more quickly noticed by birds, a sloping roof is advisable to keep food dry and fungus-free.

Seed mixes can be customized to suit the birds in your locality, but they usually include red and white proso millet, sunflower seeds and kernels, and peanut kernels. Include a small amount of grit in the mix, such as sand, to aid in the digestion of seeds.

Kitchen scraps are an asset to garden birds. Suet and other solid animal fats provide an energy-rich resource, but should be offered only when temperatures are under 70 degrees Fahrenheit (21°C), lest they go rancid. Many birds appreciate crumbs, crusts, raw dough, cooked potato, and dried-up cheese; even spaghetti, coconut, and fruit jelly are consumed. Fresh fruit, such as orange slices and grapes, are often popular during summer, when birds may find seed mixes less enticing.

THE FOUNTAIN OF LIFE

A water source is crucial to attract wildlife to the garden. It doesn't have to be large and elaborate— even a small birdbath or a rain barrel is an economical way to introduce a water feature into the garden. Ponds can serve as a visual focus, and you can create a cool and tranquil haven for people and animals with a well-planned pond area. Since the sound of running water is particularly alluring to birds, consider installing a simple water-drip device or fountain in conjunction with a small pool. Amphibians, such as frogs and toads, will also be attracted to a garden pond, along with colorful dragon-flies and other pond insects (see By the Pond on pp. 78–9).

WILDFLOWERS AND GRASSES
allowed to grow freely in the garden will not only provide food for insects and birds, but will also offer cover and nesting sites for birds and small mammals.

plant may serve multiple purposes for the same or different species. Significant shelter can also be offered in the form of nestboxes, butterfly shelters, and bat boxes.

A scarcity of nesting sites often limits the number of birds and mammals residing and raising young in the garden. Provide boxes for cavity-nesting bird species, such as bluebirds, and shelves for those accustomed to less enclosed nest sites, such as barn swallows. Consult a book that focuses on nesting aids for birds to determine the appropriate dimensions for particular species. In general, nest sites that are nestled up against concealing vegetation on two or three sides will win wider acceptance.

Fill the feeders regularly, keep the birdbath full, and plan a garden to provide a continuous supply of flowers, foliage, and seeds, and the local cast of wildlife characters will reward you with a garden full of daily drama.

Birdbaths serve birds' drinking and bathing needs. Gently sloping sides are essential to allow easy entry, and the maximum depth should be no greater than 3 inches (8 cm), with rock to serve as a perch. The perils from hunting house cats argues in favor of an elevated location, at least 3 feet (1 m) off the ground and 15 feet (4.5 m) away from dense, camouflaging vegetation.

NESTING SITES AND SHELTER

Plants provide two other essentials of habitat for the garden's inhabitants: first, opportunities for nesting, both in terms of nest material and safe locations for building; and second, shelter from predators and severe weather. For instance, a single, large, evergreen tree or shrub on a cold, windy day is invaluable to wildlife. Furthermore, a given

HOMES FOR THE BIRDS: BUILDING A NESTBOX

Different bird species prefer different-sized entry holes (refer to a good birding guide for details). Other key attributes to ensure success include a sloping roof with 3-inch (8 cm) eaves in front to prevent rain from collecting inside, four ⅜-inch (1 cm) drain holes on the bottom, and several ventilation holes near the top. The top should be screwed or hinged for easy opening and cleaning between occupancies.

Use exterior-grade plywood, 6-inch (15 cm) floorboards, or seasoned rough-grade lumber for a more rustic appearance. Alternatively, a hollowed-out piece of tree trunk suits certain species, such as chickadees. Use rust-proof nails; you may wish to paint the exterior in semi-solid stain, but don't apply any preservative inside. Unobtrusive colors are best: a light gray, tan, or brown. Mount nestboxes atop a post, or attach them to a pole or tree trunk.

PROTECTION *and* DIVERSITY

Animals may at first venture into your garden in search of food, water, and shelter, but they will stay only if they feel secure.

Protection from predators and environmental dangers is vital if you wish to witness such aspects of animal behavior as courting rituals and rearing young.

Cats lying in wait at feeders, at favorite hummingbird flowers, or at birdbaths is a common concern. One of the

THE WOOD DUCK'S *preference for cavities (top) allows it to breed in areas without groundcover. This summer garden (below) combines retreats for both people and wildlife.*

easiest strategies to circumvent predatory cats is to provide open space near these attractions so that the birds can see the threat. Chicken wire spread on the ground in early spring can deter cats from lying in a garden waiting for birds. The garden herb rue (*Ruta* spp.) planted in the problem area may also be effective in discouraging a troublesome cat.

Window-bashing by birds may need some ingenuity to prevent. Streamers hanging from eaves sometimes helps, and hawk silhouettes on the glass may work well, but you will have to move them around now and then.

Closing curtains is an obvious method to help deter birds hurling themselves into glass.

GARDEN DANGERS

Environmental danger also comes in the form of herbicides and pesticides. If you want butterflies in the garden, you must be ready to live with some caterpillars. A smattering of caterpillar-chewed leaves is testament to a balanced community in your garden; handpick the worst offenders into a jar of soapy water. Even the so-called environmentally friendly bacterial insecticide BT (*Bacillus thuringiensis*) can kill too many butterfly cater–

HABITAT DIVERSITY

Deliberately providing a wide range of blooming times, seed types, and foliage is a good strategy for attracting the greatest number of wildlife species. Diversity in the garden can be created simply by arranging the landscape into different watering zones. In general, the greater the diversity, the greater the array of wildlife that will be attracted. Vertical diversity is another way to look at the landscape. Low-growing, mid-height, and overhead plantings will be of more interest to wildlife than a landscape composed only of lawn and a few shade trees.

pillars. Likewise, the botanical insecticide pyrethrin can be highly destructive to praying mantises and butterflies. Tolerating less-than-perfect leaves and very carefully targeting only very troublesome insects are essential to successful wildlife gardening.

THE EDGE EFFECT

The location of various elements is very important. If a birdfeeder doesn't attract birds in one location, try another spot. It's not always easy to tell why a feeder might be successful in one place and not in another. The same type of nestbox located at different levels above the ground, even on the same tree, might attract different birds. Birdbaths and feeders in morning sun are much more likely to be popular than those in shady, dark morning locations. After a long winter night, the birds appreciate the warmth of the sun.

When planting areas are arranged according to the patterns found in nature, valuable edge habitat is created where two different habitats meet. Landscape edges offer many opportunities to reduce maintenance and improve the garden's appearance, and create interesting wildlife watching opportunities. In this transitional zone, you have the chance to see animals that prefer each of the two habitats alone, as well as others that require aspects of both habitats. For example, secretive animals, such as quail, dwell in dense shrubbery, but if provided with an adjoining area of low groundcover, will occasionally venture out to feed in the more exposed habitat.

SOME SPECIES *of plant and wildlife prefer the transitional zone between two habitats (top), and maximizing edge habitat in the garden is a good strategy. Despite their sometimes small size, salamanders (above) are often abundant on the woodland floor, where there is ample, moist leaf litter. The red squirrel (right) is the smallest tree squirrel in its range. It nests in hollow or fallen trees, or in holes in the ground.*

TREES *in the* NATURAL GARDEN

The symmetry, grandeur, and soaring stature of trees provide the garden with its visual framework.

As well as being pleasing to the eye, trees function as prodigious factories: from solar energy, water, and soil nutrients, they manufacture the leaves, buds, sap, wood fiber, roots, berries, and nuts that serve the nutritional needs of the animal kingdom in a myriad of ways. Many trees support caterpillars and other insects that attract insect-eating birds.

The structure of trees provides multiple opportunities for refuge and nesting: at different levels above the ground, concealed inside masses of foliage, secluded on hard-to-reach outer branches, nestled in crotches, and buried inside cavities. Birds such as orioles and black-headed grosbeaks are likely to nest in the uppermost parts of tall trees, while cavity nesters use the lower levels. Even with just a few trees growing in the garden, you can emulate the patterns found in natural forests: high canopy, sub-canopy of lower branches and smaller trees, as well as wood-land edge habitat.

Deciduous trees offer variety through the seasons. With careful selection, they can provide spring flowering, summer foliage and shade, fall color, and interesting winter branching—a changing visual treat throughout the year.

LIFE CYCLE OF A TREE

Habitat value evolves through the tree's life cycle of growth and decay. For example, a thriving, green fir tree is home to needle-eating caterpillars that are food for a family of warblers; these birds also nest in a dense cluster of its foliage. Fallen cones lure nut-gathering chipmunks. In its declining years, that same tree, laden with wood-eating insects, will provide food for a woodpecker. Elsewhere on the tree, the woodpecker excavates a hole to serve as a protected nest-site. As the tree dies and hollows out to become a snag, it provides refuge to small birds escaping the elements and a home for a family of raccoons. In its final years, the snag supports an enveloping, climbing vine that bears a heavy crop of fruit, lasting well after the warm-weather insects have disappeared.

To enjoy the benefits of a snag, allow old trees to stay in the garden, pruning only to avoid danger from falling branches. When branches do fall, let them decay in place or move them to a more suitable spot in the garden, where, in the process of becoming soil-enriching humus, they support fungi nibbled on by rodents and ferns that lure ground-nesting birds.

TREES TO ATTRACT WILDLIFE

Common name	Scientific name	Mature height	Attraction
serviceberries and shadbush	*Amelanchier* spp.	20–30 feet (6–9 m)	berries
dogwoods	*Cornus* spp.	20–35 feet (6–10.5 m)	fruits
hawthorns	*Crataegus* spp.	15–35 feet (4.5–10.5 m)	berries/cover
hollies	*Ilex* spp.	15–40 feet (4.5–12 m)	berries/cover
cherry, plum	*Prunus* spp.	15–40 feet (4.5–12 m)	soft fruits
mountain ashes	*Sorbus* spp.	20–30 feet (6–9 m)	berries
birches	*Betula* spp.	30–90 feet (9–27.5 m)	buds/seeds
oaks	*Quercus* spp.	40–125 feet (12–38 m)	nuts/cover
junipers and cedars	*Juniperus* spp.	10–50 feet (3–15 m)	cones/cover
pines	*Pinus* spp.	10–80 feet (3–24 m)	cones/cover
Douglas fir	*Pseudotsuga menziesii*	50–150 feet (15–45.5 m)	cones/cover

TREE PLANTINGS

When planting for the future, favor robust native species whose mature size suits the scale of your garden; avoid, for example, species that would ultimately enshroud the entire garden in midday shade. Nurture a variety of trees, selected according to the wildlife you wish to attract and realistic time expectations. For example, planting oaks for a bountiful acorn crop, requires a decade or more of patience, while some smaller trees will more quickly bear flowers and fruit.

Careful placement of trees is essential to the long-term success of the garden. In general, it is wise to plant trees toward the periphery to blend with adjacent habitat, to provide privacy, and to create convenient viewing angles from house and patio. If an existing tree is poorly placed—for example, shading an area you wish to develop as a sunny border—consider converting a liability into an asset by leaving the lower trunk to decay as a snag. Since mature trees are a precious resource, replaceable only over great time, try to design your garden around them.

Apart from their inherent habitat value in the garden, the ability of trees to diminish harsh winds and scorching sunlight renders them attractive to people and wildlife alike. Depending on their foliage density, and on their placement within the garden, trees can create a variety of shady exposures in the landscape, from deep, consistent shade on the north side of dense evergreens, to bright sun and dappled shade beneath the airier deciduous species. Since shaded woodland is one of the last places to dry out, apportioning a part of the garden to a low-lying forest glade offers a cool, moist retreat for wildlife during dry spells or drought.

A WILDLIFE APARTMENT BLOCK

Even an aged or dead tree can provide food and cover for a multitude of wildlife in the natural garden. Visitors might include, from top to bottom, a red-shafted flicker and its young; a gray squirrel; a screech owl; shelf fungi (a tasty treat for chipmunks and other rodents); and a marbled salamander, which will spend most of its life underground or secluded in rotted tree roots.

SHRUBS *in the* NATURAL GARDEN

Shrubs, like trees, provide a wide array of benefits for the gardener and wildlife alike.

While trees provide dramatic vertical layering within the garden, shrubs are the primary way the gardener can diversify the garden horizontally. They also provide food and cover for wildlife at a smaller scale, often down to ground level and closer to human dwellings and outdoor living space. Among shrubs are found some of the most fecund flower- and berry-producing plants, and some deciduous shrubs—such as highbush cranberry (*Viburnum trilobum*) and smooth sumac (*Rhus glabra*)—display the most brilliant fall color.

ALL KINDS OF COVER

Apart from their aesthetic role, shrubs also provide cover for shy animals. Remember that cover valuable to animals may defy conventions of what some consider a properly groomed suburban garden. For example, some gardeners may find a dense thicket of thorny shrubs, such as brambles (*Rubus* spp.), untidy and scratchy, but it will be very welcome to certain birds and reclusive mammals.

Just as trees create shady microclimates, shrubs provide the cover required by some smaller, shade-loving plants. In natural plant communities, these beneficiaries include young trees conditioned to grow up in cover provided by

NATIVE FRUIT-BEARING SHRUBS

- bearberries, manzanitas (*Arctostaphylos*) spp.
- chokeberries (*Aronia* spp.)
- barberries (*Berberis* spp.)
- salal (*Gaultheria shallon*)
- toyon (*Heteromeles arbutifolia*)
- junipers (*Juniperus* spp.)
- mahonias (*Mahonia* spp.)
- wax myrtles (*Myrica* spp.)
- chokecherry (*Prunus virginiana*)
- coffeeberries (*Rhamnus* spp.)
- currants, gooseberries (*Ribes* spp.)
- roses (*Rosa* spp.)
- raspberries, blackberries, brambles (*Rubus* spp.)
- elderberries (*Sambucus* spp.)
- snowberries (*Symphoricarpos* spp.)
- blueberries, huckleberries (*Vaccinium* spp.)
- viburnums, arrowwoods (*Viburnum* spp.)

WHEN DENSE AND TWIGGY, *shrubs are well suited to blending fences and buildings into the landscape (above and above right) to achieve a softened, seamless appearance. When pruning shrubs, take care not to shear off long-lasting berries and flower buds that promise a future food supply for birds (left, top, and right). The shrubs shown here, clockwise from above, are staghorn sumac (Rhus typhina), western azalea (Rhododendron occidentale), bearberry (Arctostaphylos uva-ursi), blue elderberry (Sambucus mexicana), and pink fairyduster (Calliandra eriophylla).*

shrubs, which, as a part of the process of plant succession, they will eventually replace. Utilize shrubs to nurture seedling trees in the garden, and either allow them to ascend and eventually take over, or intervene by removing the trees that grow beyond a desired height to maintain a complex, robust thicket of shrubs and saplings.

THE MANY VIRTUES OF HEDGEROWS

More so than trees, shrubs accept pruning to direct and contain their growth, and can be planted in dense congregations. Shrubs planted as hedges, or hedgerows, are an excellent means for interweaving several species of shrub into a narrow area to create a unified, vertical face of diverse, wildlife-friendly plants. You can customize the hedgerow to provide a nearly continuous progression of nutritious fruit and seed through all four seasons.

In garden design, hedgerows can function as dividers between gardens or sections within the garden. When at least 6 feet (1.8 m) high, they can help to screen the garden from strong winds. Depending on their orientation, they can provide different sun exposures; for example, in gardens with a

… let your garden be a wild place.

HERBERT RAVENEL SASS
(1884–1958), American nature writer and author

hedgerow established along a western boundary running approximately north to south, the east-facing exposure is favored by wildlife inclined to warm up in the morning sun—such as flying insects and the birds that pursue them. A curving hedgerow meandering through the garden offers the advantage of a variety of different sun exposures.

THE VALUE
of VINES

*Vines offer a wealth of food and cover for wildlife, and
can create a dramatic flush of color in season.*

The distinguishing value of vines lies in their flexible and often vigorous growth habits, by which they adapt to fill in shallow or narrow spaces, both horizontally and vertically. They can climb dramatically, and provide height and lush enclosure, as a background or a focal point to the garden. Despite their vigor, they are dependent upon structures and other plants for the physical support to climb.

METHODS OF CLIMBING

Curling, coiling, clasping, clinging, hooking, twining, and rambling—vines climb by many means. Some species cling firmly to wood or stone; others ascend by twining or clasping with tendrils around trelliswork or wire supports. A few, such as climbing roses, merely recline on a vertical feature and will benefit from being tied into place.

Twining vines climb by spiraling around a support; such vines include hardy kiwi (*Actinidia arguta*), pipe vines (*Aristolochia* spp.), American bittersweet (*Celastrus scandens*), honeysuckles (*Lonicera* spp.), Chinese wisteria (*Wisteria sinensis*), and Japanese wisteria (*W. floribunda*).

Some vines, such as grapes (*Vitis* spp.), climb by means of tendrils, which grow outward from the main stem and move in a circular motion until they make contact with a support. Tendrils will not coil around a horizontal support. Some tendril-climbing vines, such as Virginia creeper (*Parthenocissus quinquefolia*), have disk-like, adhesive holdfasts at the tips of the tendrils. In addition to ordinary tendrils, leaf stems, branches, and even flowers can serve as tendrils. Clematis vines (*Clematis* spp.), for example, wrap their leaf stems around supports.

Scrambling vines, such as blackberries (*Rubus* spp.), use thorns to hook themselves onto supports; while rooting vines, such as English ivy (*Hedera helix*), develop aerial rootlets to cling to supports.

VIRGINIA CREEPER (Parthenocissus quinquefolia) (top) will grow both as a groundcover and a vine. Many shrubs can be trained as espaliers, flat against the wall, such as this flowering quince (Chaenomeles spp.) (above). Rambling and climbing roses (left) can be used to form a pretty garden hedge.

two or three species together to create a tapestry-like wall of contrasting textures.

WILDLIFE VALUE

Vines provide food for wildlife, in the form of fruit, nectar, pollen, and leaves; as well as some shelter from predators and the elements. Vines can just as easily ascend shrubs, tree trunks, snags, and hedges, where they reinforce the leafy, twiggy habitat desired by nesting birds.

GROUNDCOVERS

Groundcovers help hold the soil in place, and if carefully selected, they can greatly reduce watering, fertilizing, and other maintenance tasks. They also provide cover for small creatures such as toads and salamanders, and for ground-dwelling birds such as the golden crowned sparrow. A few vines will grow well as groundcovers; for example, Virginia creeper and trumpet honeysuckle (*Lonicera sempervirens*). Other good groundcovers include grasses such as buffalograss (*Buchloe dactyloides*); evergreen shrubs such as bearberry (*Arctostaphylos uva-ursi*), junipers (*Juniperus* spp.), and creeping mahonia (*Mahonia repens*); and herbaceous garden flowers such as daylilies (*Hemerocallis* spp.).

DESIGNING FOR VINES

In garden design, vines complement architecture, softening sharp lines and angles and helping structures to blend into the landscape. Use them to disguise ordinary fences, walls, and trelliswork, and to dress up expanses of dark metalwork. In small gardens where there is not enough space for a hedge, vines provide privacy with a blanket of vegetation. Vines will divide the garden neatly into different areas, or simply provide shade. Versatile, deciduous vines trained atop a pergola filter the hot sun.

Vines such as Virginia creeper add a flare of fall color to gray tree bark and dark-needled conifers. Try growing

WILD GINGER (Asarum caudatum) (above) is an excellent groundcover for moist soil in shaded woodland gardens. Red passion flower (Passiflora coccinea) (left) and California pipevine (Aristolochia californica) (below) have unusual flowers.

VINES FOR FALL COLOR

- trumpet creeper (*Campsis radicans*)
- clematis (*Clematis spp.*)
- Virginia creeper (*Parthenocissus quinquefolia*)
- *greenbriers (*Smilax* spp.)
- trumpet honeysuckle (*Lonicera sempervirens*)
- *wild grape (*Vitis* spp.)

NON-NATIVE VINES WITH WILDLIFE HABITAT VALUE

- bougainvilleas (*Bougainvillea* spp.)
- common hop (*Humulus lupulus*)
- morning-glories, moonflowers (*Ipomoea* spp.)
- *Boston ivy (*Parthenocissus tricuspidata*)
- *passion flower (*Passiflora* spp.)
- jasmine (*Jasminum* spp.)
- *climbing roses (*Rosa* cvrs)
- potato vine (*Solanum jasminoides*)

*All vines provide cover for wildlife; these vines offer fruit.

HERBACEOUS GARDEN FLOWERS

Much of what gardeners find appealing in the form, color, and scent of flowers evolved to lure pollinators.

Herbaceous flowers are justly renowned for their prolific flowering, which simultaneously pleases the eye of the gardener and, via bountiful pollen and nectar, satisfies the food needs of butterflies, hummingbirds, and bees. Encompassing great natural and cultivated variety, these plants deliver character and vivid color even to small areas in a short period of time. Many species and cultivars serve equally well as pocket plantings, in containers, or in large, formal borders, yielding their beauty and benefits at close quarters, by paths and patios, or dispersed in casual meadows for romantic drifts of seasonal color. They are available in a range of heights and forms—from petite and single-blossomed plants to tall, impressive spikes—and there are attributes to suit any taste. Refer to The Back-yard Habitat, starting on p. 100, for more details.

FLOWERS TO LURE POLLINATORS

Flowers dependent on animals for pollination must advertise the availability of their food resource—nectar and/or the protein-rich pollen itself. Bright colors help blossoms to stand out and gain the attention of pollinators. Glossy, satiny, or velvety textures also enhance a flower's attractiveness. Sometimes, special markings, such as the speckled throat of the foxglove blossom (*Digitalis* spp.), help pollinators to zero in on their sugary target and bring them into contact with sticky pollen once they've settled on the landing strip. Occasionally, this advertisement is invisible to people: in certain cinque-foils of the genus *Potentilla*, an ultraviolet-light-reflecting center provides a target for honeybees, who cannot detect color but can see the patterns in such flowers.

The key to attracting and enjoying consistent attendance by butterflies, hummingbirds, and insects is a garden that features a sequence of flowers. Early-spring bulbs will lure the earliest bees, while a little later, the first perennial will

TRY NOT TO USE *your perennial border or meadow as a cutting garden, since you risk denying pollinators, such as the queen butterfly (top), the full benefit of this resource. Irises (left) are usually pollinated by bees, although hummingbirds rob the nectar without achieving pollination.*

bloom to welcome arriving hummingbirds. From late spring to the peak of plant growth and blooming time in midsummer, herbaceous plants join forces with flowering woody plants to serve pollinators during their season of fervent feeding and procreation. By fall and until decisive frost, blooming annuals and perennials play a critical role in providing food for lingering butterflies, beetles, and bees. By late fall and into the winter, these late bloomers pay out further

PLANT WILDFLOWERS *for a splash of color in the garden (left and above) that's accessible not only to pollinators and other wildlife but also to people. Deadheading encourages some annuals to bloom repeatedly.*

dividends in the way of seed for finches and other birds. In mild winter climates, annuals planted in fall provide floral color even when the sun is low in the sky.

North American pollinators readily accept cultivars and comparable flowers—such as daisies and roses—originating from other continents, as well as native ones. In the natural garden, however, gardeners can play a role in increasing the populations of wildflower species that are disappearing in the face of human encroachment on habitat. Since wildflowers are frequently fussier than tried-and-true commercial cultivars, the key to success in growing them lies in matching a selection of wildflowers to the climate, soil, and conditions in your garden. At the same time, consider the needs of fauna in and surrounding your garden.

PLANTING FLOWERS

Whenever possible, plant those wildflower species and varieties native to your region. They are very likely better adapted to your climate, and reduce the possibility that cross-pollination by hummingbirds or bees, who may frequent wild areas as well as your garden, will dilute the uniqueness and integrity of this local stock. If your garden adjoins a wild area, refrain from planting aggressive non-natives, which might seed into and compromise the existing native plant community.

FLOWER FAMILIES TO ATTRACT WILDLIFE

The following is a listing of genera, by family, that contain wildlife-pleasing species.

- Asteraceae, sunflowers (*Aster, Helianthus, Gaillardia, Chrysanthemum, Coreopsis, Liatris, Rudbeckia, Solidago*)
- Lamiaceae, mints (*Lavandula, Mentha, Monarda, Salvia, Rosmarinus*)
- Scrophulariaceae, figworts (*Penstemon, Antirrhinum, Digitalis, Mimulus*)
- Cruciferae, crucifers (*Lobularia, Matthiola*)
- Leguminosae, peas (*Lupinus*)
- Lobeliaceae, lobelias (*Lobelia*)
- Violaceae, violets (*Viola*)
- Verbenaceae, verbenas (*Verbena*)
- Onagraceae, evening primroses (*Clarkia, Oenothera, Zauschneria*)
- Papaveraceae, poppies (*Eschscholzia, Papaver*)
- Polygonaceae, buckwheats (*Eriogonum, Polygonum*)
- Ranunculaceae, buttercups (*Anemone, Aquilegia, Delphinium, Ranunculus*)

GRASSES *and* WEEDS *in* *the* NATURAL GARDEN

Free-growing grasses do not demand the same intensive

maintenance as mowed turf grasses, and offer far greater variety.

Ornamental and native grasses, along with grass-like sedges and rushes, comprise the key element in the creation of meadows and prairie-like habitats. They also serve well at the edges of streams and ponds and as a patchy, informal groundcover.

The fine, delicate textures of grass leaves, flowers, and seed heads, waving to the tempo of the wind, are a welcome contrast to the boldness and solidity of other herbaceous and woody plants. Their subtle, modulated colors include creamy and silvery variegation and shades of purple, bronze, orange, and red, especially in fall and winter. While many are low-growing, others become impressive upright or arching specimens. Some, such as giant wild rye (*Elymus condensatus*), are tall enough to serve as windbreaks.

Sod-forming grasses, such as Kentucky bluegrass (*Poa pratensis*) and buffalograss (*Buchloe dactyloides*), spread either by above- or below-ground runners, and are generally found in moist sites. Bunchgrasses don't spread, but form clumps and are generally more common in dry locations. Many native and introduced bunchgrasses are popular ornamental plants for gardens, such as blue grama (*Bouteloua gracilis*), Indian grass (*Sorghastrum nutans*), Indian rice grass (*Oryzopsis hymenoides*), little and big bluestem (*Schizachyrium scoparium* and *Andropogon gerardii*), and sideoats grama (*Bouteloua curtipendula*). (Refer also to The Backyard Habitat, starting on p. 100, for details about grasses for your region.)

ORNAMENTAL GRASSES *such as these Miscanthus cultivars (above) provide variety in form and color in the garden. Plant grasses with wildflowers to re-create natural habitat for wildlife (above right and right).*

THE BENEFITS OF GRASSES AND WEEDS

The deep, fibrous roots of grasses help to forestall erosion by anchoring soil, and most are well adapted to poor soil conditions. Most native grasses are resilient in the face of foliage-browsing insects and mammals, provided their numbers are moderate. Many tolerate dry, exposed conditions, while others thrive in damp, shady situations.

So-called "weeds"—those tough, hard-to-eradicate herbaceous plants with modest appeal as garden specimens— produce crops of seeds with substantial food value for wildlife, especially birds. Weeds are easily tolerated when merged with grasses and showy wildflowers, though you may choose to omit the nutritious but allergy-inducing ragweed (*Ambrosia* spp.). Compose your meadow to suit your taste, the local character of your site, and the specific wildlife you aspire to attract by devising your own customized planting plan, as opposed to relying on the widely available field-in-a-can seed assortments. Be sure to avoid planting two invasive perennial grasses, Bermuda grass (*Cynodon dactylon*) and quack grass (*Agropyron repens*), lest they proliferate to the exclusion of other useful attractive species. And take care to contain these aggressive grasses: pampas grass (*Cortaderia* spp.), common reed (*Phragmites australis*), and ribbon grass (*Phalaris arundinacea* var. *picta*).

GRASS MEADOWS AND WILDLIFE

What appears to be a graceful, flower-studded plain, rippling in the breeze, is actually a complex habitat of densely ordered plants, with vertical definition and a range of wildlife habitat opportunities, or niches. A forest in itself, a botanically diverse meadow supports a multitude of beetles, bugs, flying insects, and spiders. Reptiles such as the eastern box turtle hunt for earthworms, beetles, caterpillars, and berries in damp meadows, and even some amphibians take advantage of a grassland's abundant summertime food. Naturally, many insect-eating birds are drawn to grassy habitat to feed and raise their young.

NON-INVASIVE GRASSES AND WEEDS FOR THE NATURAL GARDEN

Grasses *denotes non-natives
- bluestems, beardgrass (*Andropogon* spp.)
- gramas (*Bouteloua* spp.)
- reed grasses (*Calamagrostis* spp. & cvrs)
- hairgrasses (*Deschampsia* spp. & cvrs)
- wild rye, Pacific dune grass (*Elymus* spp.)
- plume grasses (*Erianthus* spp.)
- fescues (*Festuca* spp. & cvrs)
- *blue oat grass (*Helictotrichon sempervirens*)
- *silver grasses, maiden grass (*Miscanthus* spp. & cvrs)
- muhly grasses, deer grasses (*Muhlenbergia* spp.)
- switch grass, deer tongue grass (*Panicum* spp.)

- little bluestem (*Schizachyrium scoparium*)
- Indian grass (*Sorghastrum nutans*)
- dropseeds (*Sporobolus* spp.)
- feather grasses, needle grasses (*Stipa* spp.)
- gamma grasses (*Tripsicum* spp.)

Weeds
- turkey mullein (*Eremocarpus setigerus*)
- red-stemmed filaree (*Erodium cicutarium*)
- bush clovers (*Lespedeza* spp.)
- common tarweed (*Madia elegans*)
- pokeweed (*Phytolacca americana*)

PLANTS *for the* BACKYARD POND

A water source, especially running water, is the key to luring birds and mammals to the garden.

Watery habitats in natural ecosystems are vitally important. The sudden splash of a frog, the darting of a fish-eating bird, the flash of a dragonfly: water appeals strongly to wildlife, which will congregate by ponds and streams for drinking, feeding, and bathing. In the case of amphibians, a clean, watery environment is necessary for reproduction as well.

THE CENTERPIECE OF THE GARDEN

Ponds are well suited to serve as the centerpiece of a garden: their shimmering, sky-reflecting surface opens up a busy landscape and they complement all variety of plant forms.

In addition to the soothing sound of water, a pond will moderate the extremes of heat through surface evaporation, and create a microclimatic

Nights of watching … will let us into some secrets about the ponds and fields that the sun … will never know.

DALLAS LORE SHARP (1870–1929), American author, naturalist, and educator

effect enhanced by the flow of air over the pool—like in an oasis. The sound of running water can also be used to buffer the harsh sounds of urban life, such as vehicular traffic.

Depth is important: frogs prefer depths between 6 and 24 inches (15 and 60 cm), along with sloping walls that permit adults to exit; while waterbirds favor ponds with an average depth of about 18 inches (45 cm).

To reap the full benefit of a garden pond, gardeners must plan carefully and provide a clean source of water. Pollution from septic tanks and excess fertilizer from farmyards and gardens interfere with the natural chemistry of a body of water. Pesticide residues, swimming-pool chlorine, and petroleum-tainted runoff from highways all have the potential to poison aquatic life.

POND ZONES AND PLANTS

Different plants are suited to different zones in the pond: for example, submerged plants such as water milfoil (*Myriophyllum* spp.); floating plants, such as water fern (*Azolla* spp.) and duckweed (*Lemna* spp.); emergent plants, with their roots on the bottom, such as waterlilies; and plants for the pond margin, such as cattails (*Typha* spp.), pickerel weed (*Pontederia cordata*), and sedges (*Carex* spp.).

If waterfowl are a priority, it's best to grow some resilient native plants that particularly suit their tastes; these include pondweed (*Potamogeton* spp.), wild celery (*Vallisneria americana*), and wild rice (*Zizania aquatica*)

If waterfowl are not present, you can grow interesting foliage plants such as arrow-

GARDEN PONDS *A natural stream running through a garden is worth treasuring and enhancing, but the stiller water of a pond (above and right) supports a more diverse array of water plants and provides a greater variety of niches for wildlife.*

head (*Sagittaria* spp.) and the native waterlilies. The latter are of particular value, not only for their showy flowers and long bloom season, but also for the variety of habitats they offer the smaller pond–dwelling animals. Buds and blossoms provide convenient perches for dragonflies; their bold, floating leaves form sunny platforms for frogs. Beneath the surface, some very small pond animals, such as rotifers, hydra, and insect larvae, attach to their stems. These wide-leafed plants also benefit wildlife by partially shading the pond water, preventing overheating during intense summer sun-shine, and by providing seclusion for fishes, tadpoles, and salamanders. Prime possibilities include the yellow pond lily (*Nuphar advena*) and the fragrant waterlily (*Nymphaea odorata*).

As well as introducing oxygen into the water, water-lilies achieve two other important goals: excess carbon dioxide is removed and the sun's energy is utilized to construct the plant tissue that feeds wildlife. Plants of particular merit in aerating the pond include the bladderwort (*Utricularia vulgaris*), water milfoil, water pennywort (*Hydrocotyle* spp.), and water shield (*Brasenia schreberi*).

GETTING THE MOST FROM YOUR POND

- Emulate nature by positioning the pond in lush vegetation.
- Ensure the pond is not entirely visible from any one vantage point, to create an air of mystery and to invite discovery.
- Situate the pond in a low point in the landscape, where water would naturally collect.
- Choose plants such as yellow flag (*Iris pseudacorus*), which will provide a contrasting vertical line in its growth habit.
- Lush foliage hanging over or creeping up to the water will soften hard corners or edges. Good pond-edge plants include wild iris (*Iris versicolor*), marsh marigold (*Caltha palustris*), and spike rush (*Eleocharis montevidensis*).
- To create harmony in the garden, use stones to edge the pond only if they are present elsewhere in the garden.
- Don't allow bare concrete or exposed plastic liner to show, or cover them with ground-hugging plants such as marsh marigold (*Caltha palustris*) or one of the sedges (*Carex* spp.).

... all those little fiddles in the grass, all those cricket pipes, those delicate flutes, are they not lovely beyond words when heard in midsummer on a moonlight night?

HENRY BESTON (1888–1968),
American writer and editor

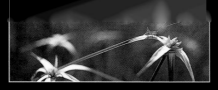

CHAPTER THREE

INHABITANTS *and*
PASSERSBY

FROM TREETOP
to BURROW

Natural gardening at its best provides welcome

habitat for a diverse community of wild creatures.

Gardeners attuned to the well-oiled machinery of natural ecosystems can design plantings that are not simply an amenity for people. They can compose habitat essentials for birds and mammals, even amphibians, reptiles, and insects, to create a garden that both satisfies the eye and attracts and supports these animals.

In the flush moments of early-morning feeding, during migrations, or as night-loving animals emerge at dusk, the natural garden may resemble a busy stage-set, as characterful as a Dickens novel, humming with the pulse and spontaneity of the wild. In this thriving sanctuary, far removed from the manicured yard adorned with picture-perfect, exotic flowers, you may be witness to behaviors as poignant as the feeding of young, as absorbing as a fierce dispute over territory, or as dramatic as predators stalking and capturing their prey.

BARN SWALLOWS *(top) are swift and graceful flyers. Black-eyed Susans* (Rudbeckia hirta), *(right) provide a good source of nectar for honeybees and butterflies.*

PLANNING YOUR SITE

A garden comprising a variety of plant life provides a mosaic of opportunities—called niches by ecologists—for different species of wildlife. Throughout the year, hundreds of insect species and scores of bird species will frequent the garden planned for *diversity*. Of key importance here are the food resources offered by trees, shrubs, and herbaceous plants. Nesting sites and shelter from the elements and predators are of equal importance. For example, a tree swallow may be drawn to a backyard to feed upon a wealth of flying insects over a pond and patch of meadow, but will reside in the garden only if a suitable hollow tree or nesting box is available.

All nature wears one

universal grin.

HENRY FIELDING
(1707–54),
English writer.

Many small mammals require secluded recesses or grassy cover to escape predators, such as the hungry hawk or owl. In some cases, an animal's needs vary through its life cycle: in many butterflies, the caterpillars depend on particular foliage, while the adults prefer the nectar of certain flowers. And don't neglect water: it's a magnet for all types of creature, particularly where it's scarce, as in the desert and during frozen northern winters.

PLANTING TIPS TO ATTRACT WILDLIFE

• Plan the garden around a backbone of native and non-invasive adapted plants. Re-creating the attractive qualities of healthy plant communities—such as species diversity, toughness, and the provision of reliable resources—gives you the best chance of meeting the specific needs of native fauna in the backyard habitat.
• Take care to preserve tall trees, since they yield the most dividends in the way of food, cover, and habitat complexity. In their absence, think long-term and plant a grove for the future to restore some of the forest sacrificed to human needs.
• If your yard is too small for substantial trees, create vertical niches with vines. Plant two or three different species together to make a tapestry of habitat.
• Plant for a seasonal sequence of flowering and fruiting to provide wildlife with food sources throughout the year.

You can learn much from natural ecosystems when designing your garden for wildlife. Establishing a vertical layering of vegetation, from low groundcover to high canopy, like that found in nature, can be as important as plant variety. By graduating garden plants from short ones in the foreground to progressively taller ones farther away, you'll optimize your viewing opportunities in all the different layers.

Horizontal habitat diversity is also important. Different habitats will attract different species, of course, and edges, where different habitats converge, attract still others. The transition between forest and shrubbery, the zone straddling meadow and thicket —each of these harbors more species than isolated patches of any one habitat alone, because they can satisfy a greater variety of animal needs. For example, certain flycatchers favor high, exposed perches at the forest edge where they can hunt insects that fly over a meadow. Both habitats are required—one for the food, one for the perch—

for this species to prosper along with strictly woodland and meadow bird species.

Keep in mind that a garden both layered vertically and diversified horizontally must be organized in a natural way to be effective. For example, a stream winding through a variety of low, leafy plantings is more attractive to many animals than one that flows within an open expanse of patio or lawn. One reason for this is that many animals are as concerned as we are with personal safety. A handsome but shy woodland bird that feels secure bathing in a stream enclosed by shrubbery beneath oak trees would not feel the same security about venturing into the more exposed, and hence, more dangerous, mid-lawn waterway.

A GARDEN *planned for diversity (top) creates different microhabitats and a variety of niches for wildlife. The monarch (above right) is one of the best-known and most widespread of North American butterflies, while the Olympic marmot (right) is found only in the Olympic Mountains of Washington.*

BUTTERFLIES *and* MOTHS

The power of flight enables butterflies and moths intimate and useful contact with flowering plants.

Butterflies and moths belong to the same taxonomic order, the Lepidoptera. As much as the many species vary in size and color, they all share a four-part wing structure: that is, a pair of forewings and a pair of hindwings. At rest, butterflies often close their wings together and hold them upright, showing only dull, camouflage-colored underwings—a strategy for avoiding detection by predators. Many moths, on the other hand, achieve the same protection with cryptically colored (camouflaged) forewings that they hold roof-like over their bodies to conceal themselves against tree bark.

Another obvious difference between the two groups is that moth antennae are often elaborately plumed, or feathery, in comparison to the simpler, knobbed antennae of butterflies.

MULTICOLORED WINGS

The resplendent colors of many butterfly and some moth wings are formed by a tapestry of millions of delicate, microscopic scales arranged atop the otherwise clear wing membrane. Each species sports its own distinctive wing-color pattern, and the coloration of females may differ from that of males—a distinction that facilitates courtship.

Butterflies and moths are able to fly only when their cold-blooded bodies are warm enough—butterfly wings, laid open to the warmth of the sun, act as small-scale solar panels in cool weather. Moths differ from their sun-loving relatives in that most are night-flyers, so they must shiver to warm up their flight engines.

Flowers provide an energy source for adults, and they, in turn, transfer pollen—usually accidentally—thus assisting the flowers in cross-pollination. Butterflies and moths imbibe nectar (and sometimes water) via a proboscis, a long, coiled tube situated on the head. The rear pair of their six legs is equipped with taste organs that assess the suitability of a food source before feeding.

A VARIED LIFE

On the path from fertilized egg to breeding adult, insects typically pass through a series of physical stages called instars, each separated by a molt of their hard, outer skeleton (the exoskeleton). When the transformation between stages includes a crawling larval stage dramatically different from the adult stage, it is called complete metamorphosis, a process most readily appreciated in butterflies and moths.

Shortly following mating with a male of her species, the female seeks out appropriate host plants on which to deposit her eggs. A butterfly hovering deliberately over foliage could be searching for suitable host plants in the garden using her smell and taste organs. The egg is small—typically, no larger than the

THE IO MOTH *(top) sports boldly colored hindwings that are visible only in flight. Many of the dazzling hues of butterfly wings, especially blues and greens (left), are created by iridescent scales. This close-up of callippe fritillary wings (above) shows the minute scales.*

THE LIFE CYCLE

◀ **❶** *A single female can produce eggs in the hundreds, yet each is deposited singly, either on or under a leaf of the host plant.*

❷ ▶ *The active and voracious larva, the caterpillar, emerges from the immobile egg.*

❸ ▶ *The pupa encases the caterpillar that created it, and rests in a protected spot while the tissue of the larva inside dissolves and reorganizes into a winged adult.*

◀ **❹** *After a week or two, the adult butterfly— here, the gulf fritillary— breaks the pupa case and crawls out.*

head of a pin. The hard, protective shell, initially pale, darkens as the embryo grows.

After emerging from the egg, the caterpillar's purpose is relentless consumption of the foliage of the host plant. With a suitable supply of food, the larva can outgrow and shed its exoskeleton four times in three to six weeks, at which point, it is ready to molt into a pupa.

In butterflies, the pupa, or chrysalis, hangs on a narrow stalk from a plant stem or beneath a leaf. You can identify the chrysalis of many species from its shape and color.

Moth pupae are usually wrapped in silk cocoons. In many species, the pupae are hidden within a plant stem, in leaf litter, or underground in an earthen cell. Some species overwinter as dormant pupae in such locations.

Over a period of hours after emerging from the pupa case, the adult butterfly or moth unfolds its still-tender wings, pumping them up with fluid to achieve their glory and air-worthiness. New emergents may appear almost aimless in their flickering travels in the garden, but their life continues in earnest as the location of nectar and a mate become their main occupations.

BEST BETS
Plants that most reliably attract butterflies

- dogbanes (*Apocynum* spp.)
- *pipevines (*Aristolochia* spp.)
- wormwood (*Artemisia dracunculus*)
- milkweeds (*Asclepias* spp.)
- asters (*Aster* spp.)
- butterfly bush (*Buddleia* spp.)
- California lilacs (*Ceanothus* spp.)
- bull thistle (*Cirsium vulgare*)
- buckwheats (*Eriogonum* spp.)
- mints (*Mentha* spp.)
- black-eyed Susan (*Rudbeckia hirta*)
- goldenrods (*Solidago* spp.)
- *nettles (*Urtica* spp.)
- violets (*Viola* spp.)

Consult The Backyard Habitat (starting on p. 100) and a good butterfly field guide before planting.

* These species provide only larval food.

AT NEARLY 6 INCHES *(15 cm) wide, the giant swallowtail is one of the largest butterflies in North America.*

PLANT LOVERS

Insects have evolved extremely varied ways of life

and diets, yet most are intimately associated with plants,

often as consumers of sap or foliage.

In order to avoid detection by predators, many insects are colored in hues of green or brown that match the leaves or bark of their favored habitat. To reach adult stage, plant-eating insects undergo incomplete, or simple, metamorphosis. After they have hatched from an egg, the nymphs grow incrementally, shedding their exoskeleton several times until they achieve their adult characteristics such as the ability to reproduce and the formation of wings.

KATYDIDS AND KIN

Katydids (also known as long-horned grasshoppers) are at home amidst the dense foliage of deciduous trees and shrubs. Weak flyers, their camou-flaging wings see important service in the production of the sound males use to communicate with potential mates. Customarily, it is only the male that sings, but in one species, the broad-winged katydid, the female responds in kind. When singing, they raise up the base of the forewings and rub a ridge of horny bumps one against the other, much like a fingernail on a comb. Only the species called the true katydid produces the famous ditty "Katy did" and, less frequently, "Katy-didn't". The songs of other species consist of sharp, assertive clicks and buzzes, delivered in recognizable and oft-repeated patterns.

The order Orthoptera, to which katydids belong, also includes the grasshoppers and crickets. They all share similar leaf-chewing mouthparts and prominent, often powerful hind jumping legs. As for the damage they inflict, seasonal browsing by these insects does not usually have a major impact on the appearance of the garden, especially if the broad-leafed and grassy plants they favor are varied and robust.

Many of the grasshopper species are found near the ground, such as the American bird grasshopper, which prefers low shrubbery or meadow. Grasshoppers are more likely to jump and fly than their tree-dwelling cousins, and they sometimes display boldly marked hindwings. They include species that are champion flyers, such as the migratory grasshopper, which occa-sionally appears in sudden, localized mass infestations that bedevil farmers. Their singing repertoire includes extended buzzes and crackles, some-times delivered in flight.

Another famous jumper, the familiar and widespread field cricket, utters a series of quick, three-part chirps that

CAMOUFLAGING PLOYS *Cousins of the leafhoppers, treehoppers such as this oak treehopper (top) mimic the shapes and colors of the thorns and buds of their host plants. Like most katydids, this fork-tailed bush katydid (left) has veined, leaf-like forewings, which cover most of its body and serve as camouflage in the wild.*

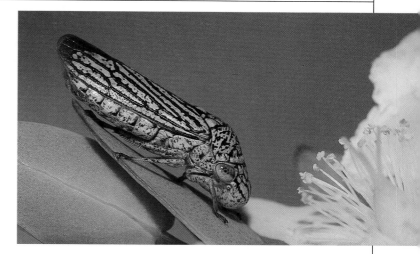

THIS SHARPSHOOTER LEAFHOPPER
(right) gets its name from its ability to leap with the speed of a flying bullet. The American bird grasshopper (below right) feeds on grasses and other herbaceous plants.

are quite musical. This species is nearly black in coloration, which is good camouflage for the shady, damp undergrowth it favors as habitat.

The snowy tree cricket, found in most of North America except the Southeast United States, is pale green to blend with its leafy habitat. Like other crickets, the tempo of its trilled chirps increases with temperature. Count the number of chirps in 15 seconds and add 37 to find the approximate temperature in degrees Fahrenheit. Although the adult snowy tree cricket is a carnivore and feeds on aphids and caterpillars—a beneficial role that the gardener will appreciate—snowy tree cricket nymphs feed on foliage, flowers, and other plant material.

JUICE LOVERS

The members of another order of insects, the Homoptera, or full-winged bugs, rely on the succulent juices of plants for their food. Among the most abundant of these bugs are leafhoppers, which favor meadows and borders and are inclined to jump suddenly when disturbed.

The largest of the Homoptera, the cicadas, have stout, rugged bodies and prominent eyes. As with other members of the order, their mouthparts are located well underneath the head. The cicadas include some species

famed for their ability, as nymphs, to feed on underground roots for long periods—in one case, for 17 years. Only then do they emerge en masse as adults to confound predators with their numbers, making it easier for them to mate and breed. The males deliver an intense droning or pulsating buzz from high perches in trees—a sonic fixture of lazy, summer days in many parts of North America.

The other major order of plant-juice opportunists, the Hemiptera, or true bugs, are sometimes called the half-winged bugs, since only the tip of the forewing is a clear membrane, while the base is leathery. Their divided wings folded across their back create a diagnostic "X" pattern, and their sucking mouthparts project forward from their heads. Among the most likely to be encountered in North American gardens is the shield-shaped, green stink bug, so named for the foul-smelling fluid it discharges as a deterrent to predators.

THE JAGGED AMBUSH BUG

The jagged ambush bug, a relative of the vegetarian milkweed bug, has adapted the same Hemiptera body form to prey on visitors to flowers. It seizes bees and other pollinating insects with its formidable forelegs, and then uses its sharp beak to paralyze them and ingest their body fluids.

THE FREQUENT FLYERS

Frequent flying enables most groups of insect to participate in the pollination of flowers, predation, and ingenious forms of parasitism.

Many adult insects have wings, but those that rely most on agile, rapid flight are the bees and wasps, the flies, and the beetles. All develop via complete metamorphosis; that is, they pass through a distinct, crawling larval stage before emerging as airborne adults.

ARCHITECTS BY INSTINCT

Although it is not native, the honeybee plays a greater role in pollination and the resulting setting of seed and fruit than any other insect in North America. Honeybees nest in hollow trees but readily accept human-made hives. Either way, they construct a water-resistant, waxen comb made up of a regular matrix of hexagonal cells. These domestic building-blocks serve as homes for larvae or as storage bins for the honey that feeds the colony when flowers and their nectar are scarce.

As with many bees and wasps, the life cycle focuses on the fertile queen, who produces all the larvae, numbering 50,000 or more. The larvae are fed primarily bee bread, a combination of honey and pollen, which the worker bees collect in basket-like indentations in their hind legs. In spring or summer, the queen bee flies out of the hive, accompanied by a swarm of workers, to found a new colony and mate with the short-lived drones. A daughter reared on "royal jelly", a highly nutritious mixture secreted by the worker bees, succeeds her mother as queen of the old hive.

The black-and-orange bumblebee is another familiar social bee. In contrast to honeybees, whose colony members live through the winter, it is only the queen bumblebee that survives to emerge in spring and search for a protected crevice in which to found a new colony.

Among the bees, the social bee species dominate in sheer numbers, but in diversity of species and ways of life, the solitary species reign supreme. Most solitary bees lay their eggs and rear their larvae in underground holes or other low, damp places where mold is a menace to the eggs and larvae. The plasterer bee protects its young by coating the cells in which they hatch and grow with a special saliva, thus forming a waterproof seal.

Excellent pollinators that don't sting, mason bees mix stone dust or clay with saliva

MANY PLANTS *rely on insects such as honeybees (top) and adult long-horned beetles (above) for pollination. Bumblebees (right) are good pollinators, but unlike honeybees, they can sting repeatedly.*

THE BALD-FACED HORNET

Among the fiercest of the predatory wasps, the bald-faced hornet patrols sunny walls in order to launch aerial assaults on flies. Hornet species construct nests of "paper" derived from chewed wood fiber. As the colonies grow, hornets enlarge their nests, layer upon layer, to create an impressive gray orb attached to a shrub or tree, often by the forest edge. The queen can be seen here on the right.

to form a cement-like building material. The female uses the material to construct a group of cylindrical cells, situated underground or camouflaged against a rock. Provisioned with honey and a single egg laid in each, the cells are then covered over in the final act of instinct-driven masonry. Growth and metamorphosis of the next generation take place in this dry, secure setting. The new adult chews its way through the durable cell wall to emerge.

WASPS AND HORNETS

While bees are nectar- and pollen-harvesting vegetarians, wasps cruise the garden to bring in fresh-killed caterpillars and other insects as food for their larvae.

Like bald-faced hornets, yellow jacket wasps build a "paper" nest, but conceal it underground or in a crevice. Accidentally disturbing an occupied nest is perilous, since wasps will defend their homes en masse, and individuals sting repeatedly.

In fall, hornets and yellow jackets feed upon overripe fruit, nectar, and, occasionally, hummingbird feeders. They can be an annoying but not especially dangerous presence in the garden.

Other wasps include the less ferocious paper wasps, who build small colonies consisting of one layer of cells, often under the eaves of buildings; and the solitary potter wasps, who build their delicately curved, single-cell clay chambers on twigs.

INGENIOUS PARASITES

While the colony-nesting wasps control the proliferation of many plant eaters through direct predation, the braconid, chalcid, and ichneumon wasps parasitize the still-living caterpillars. The female wasp locates the larvae of its host species and lays eggs upon their backs. The wasp larvae hatch and then burrow into their host, feeding upon and eventually killing it, but only after the young wasps are ready to pupate. Many of

these wasps, including the intriguing hyperparasites that lay eggs on the larvae of other parasitic wasps, are so small as to be barely detectable by the unaided eye.

The beneficial role these parasites play in controlling such pests as the cutworm and the fall armyworm is of increasing interest to farmers and gardeners. For example, a chalcid wasp has been deliberately introduced from Europe to help in the control of the non-native gypsy moth, which is a defoliator of Eastern forests. Parasitic wasps are commercially available in pupal form—sometimes being provided as small dots on a card—and can prove useful inside greenhouses and on farms, but probably disperse too quickly to be of great use in the backyard garden. However, gardeners can encourage local populations of these beneficial wasps simply by planting herbaceous flowers for the wasps to feed on, such as dill, parsley, yarrows *(Achillea* spp.), and tansy *(Tanacetum vulgare).*

BENEFICIAL INSECTS *Because its larvae prey on pests such as aphids and scale insects, the American hover fly (below) is considered highly beneficial in the garden. Ladybugs also prey on aphids and other garden pests.*

WIGGLERS *and* CRAWLERS

A host of inconspicuous creatures works below the gardener's feet to create a more favorable environment for plants.

One mainstay of garden ecology is the earthworm. In a healthy garden, earthworms consume great quantities of decaying leaves and grass, converting this plant detritus into rich humus. Their quiet, extensive burrowing creates a myriad of passageways through which air can circulate more easily around root systems, where it is needed to ensure healthy microbiological activity in the soil. This loosening of the soil also allows water and water-borne nutrients to be absorbed more readily into growing roots.

CONSUMERS OF DECAY

Two common types of earthworm frequent gardens: the field worm and the so-called night crawler. To encourage these hidden heroes of the garden, avoid excessive use of chemical fertilizers. Nutrient-rich compost is a much more compatible additive to garden soil and is also favored by the earthworm's allies in soil

improvement: millipedes and sow bugs, as well as minute springtails and soil mites, which also help to break down decaying plant material.

UNDERGROUND INSECTS

Like the burrowing of earthworms, the subterranean tunneling of ants assists in aerating the soil. The little black ant, a widespread species that enjoys a wide-ranging diet, maintains its subsurface

colonies near forest edges and by human habitations. Other ant species follow much more specialized diets: the red ant collects nectar and also honeydew, the sweet secretion of aphids. These ants crawl up plants to reach open blossoms and feeding aphids, returning to their underground nests with their sweet bounty.

The crawling larvae of many of the flying insects are

THE ELEPHANT STAG BEETLE
(top) defends itself with its fierce-looking pincers. It feeds on nectar and plant juices. The night crawler earthworm (right) burrows to the surface to feed upon damp leaf litter when the sun is down.

THIS SIX-SPOTTED *green tiger beetle (above) races across rocky screes and along dirt paths to seize passing ants and other insects. The large desert tarantula (left) has nocturnal hunting habits.*

adapted to dwelling underground. May and June beetles spend their larval stage in soil, chewing on plant roots. Other larvae are sophisticated predators. The tiger beetle larva hides in a hole to grab unsuspecting ants and spiders. The voracious ant lion larva pursues a similar strategy, but waits at the bottom of a cone-like depression in sand, into which hapless prey tumble.

DEADWOOD RECYCLERS

Other insect larvae, including many species of beetle, dwell out of sight in decaying tree stumps and fallen limbs, where they undertake the difficult

THE BLACK CARPENTER ANT
(below) feeds on other insects, honey-dew, and fruit juice. It builds its nests inside wood but is not a wood eater.

process of digesting wood cellulose, or fiber. Their wood-boring habits leave elaborate patterns beneath the bark of deadwood, testifying to their useful role in breaking down this resource into natural fertilizer. Woodpecker species that hunt for these wood borers are also major beneficiaries when deadwood is allowed to remain in the garden.

TERRESTRIAL PREDATORS

Whether shady, leafy, or sunny, the ground level of the garden provides an arena for a variety of active predators. One commonly occurring group of insect predators, the ground beetles, plays a beneficial role by avidly hunting caterpillars and even slugs and snails. Both the larva and the adult of the fiery searcher, a ground beetle found throughout North America, frequent gardens and cultivated fields and even climb trees in pursuit of their prey. To encourage it and other beneficial beetles such as tiger beetles, provide permanent refuge with beds of perennials or low shrubs.

EIGHT-LEGGED GROUND DWELLERS

Spiders also number among the important ground-level predators. Rather than tending a web, the thin-legged wolf spider patrols a territory out in the open, basking in the sun at the meadow's edge and chasing down its insect prey. Typical of most spiders, it injects its victims with paralyzing venom and then sucks them dry until only the exoskeleton remains.

The large desert tarantula of the Southwest is hairy with a heavy build, but contrary to reputation, its venom is no worse than the sting of a wasp.

The trap door spider of southern California perfects the hide-and-surprise strategy. Atop its underground chamber, this enterprising spider fashions debris, soil, and silk from its spinneret into a hinged "door", which it flings open to capture unsuspecting insects and even other spiders.

The daddy longlegs, which is technically not a spider, has very long, thin legs that arch over its small body. It hunts small insects and imbibes plant juices, mainly at night.

BY *the* POND

A garden pond or stream will do more to attract wild creatures than any other single improvement.

Beyond its aesthetic value, water is essential to life. A variety of insects and birds finds food in or around water, and virtually all creatures, including some otherwise secretive birds and mammals, are drawn to ponds and streams to drink.

WATER LOVERS

Amphibians are among the easiest water-loving animals to oblige. Certain frogs can prosper and breed in very small pools—the size of a large kitchen sink will suffice.

Details of the life cycle vary with the species, but breeding customarily

THE PACIFIC TREE FROG *(below) is less arboreal than many of its relatives, and can be found in low shrublands and damp meadows. Dragonflies (below right) are beneficial predators that devour huge numbers of mosquitoes.*

commences when males vocalize on spring nights to attract females, forming loud choruses on occasion. After mating, females lay fertilized eggs in water; the eggs stay submerged in a mass, some-times attached to vegetation. Tadpoles hatch from the eggs, and capture insect larvae and other aquatic life. They typically spend several months underwater, gradually trans-forming themselves into four-legged adults lacking tails. Strong hind legs power energetic leaps either in pursuit of prey or to avoid predators, while webbed feet enhance their prowess as swimmers.

If your new garden pond has not yet attracted any migrating frogs, you can establish a population by transferring a small number of eggs from nearby natural ponds. Many species abound in the rainier eastern and central parts of North America, including the pickerel frog, the green frog,

AFTER WINTERING *in warmer climes, Canada geese return each year in spring to their ancestral breeding grounds.*

the northern and southern cricket frog, and the spring peeper, well known for its reedy, nighttime chorus. The red-legged frog may show up in West Coast water gardens, along with the especially musical Pacific tree frog.

Toads, too, lay eggs in water, but the tadpoles develop much more rapidly, and even a temporary pool in a depression will suffice. Adult toads lead largely terrestrial lives, preferring damp, shady recesses in the garden and by the house. Like all amphib-ians, they need some contact with moisture to prevent drying out. Toads have neck glands that secrete a poison to deter would-be predators.

WARM, SPRINGTIME RAIN *will often see the spotted salamander (right) emerge to breed in woodland ponds. The iridescent colors of male wood ducks (below right) are a visual delight on woodland ponds and streams.*

A toad's typical diet includes major garden pests, such as cutworms, armyworms, slugs, and snails, as well as the beneficial earthworm.

The Woodhouse's toad, found in the central United States, adapts well to gardens. Primarily a nocturnal insect hunter, it snaps at flying bugs attracted to house lights. During the day, it rests hidden in vegetation or retreats into its burrow. The spadefoot toads, of which there are several North American species, lack the poison glands of the true toads, but excel at accelerated breeding: a new generation emerges from short-lived puddles in less than two weeks.

FEATHERED POND DWELLERS

Potential avian visitors to larger ponds include some avid vegetarians, including the Canada goose, the American coot, and various ducks. The widespread mallard is the duck most likely to visit your pond; it will dabble at the surface or browse succulent underwater plants, occasionally taking fish and frog eggs and aquatic insects as well.

POND INSECTS

Well-aerated water makes it possible for animals that absorb oxygen from the water to breathe and thrive. These include animals with gills (fish, tadpoles, and salamander larvae) and a multitude of aquatic insects that breathe through their skin. Among the most common insect

larvae that use this oxygen are the dragonfly nymphs. After feeding voraciously on the larvae of mosquitoes and other pond dwellers, they emerge as agile, fast-flying adults. Patrolling ponds and streams for adult mosquitoes, midges, and flies, they have been clocked at speeds up to 60 mph (100 kph). The female lays eggs on floating leaves and moist rocks.

Other predator insects operate at the very surface of the water itself. Water striders glide across almost every pond or stream, relying on the surface tension of the water for support. Like the land-dwelling spiders they superficially resemble (they are actually related to the plant-sucking bugs), water

striders sense the vibrations of any small, hapless insect caught on the surface and scoot over to give a paralyzing kiss.

Groups of small, black, oval whirligig beetles, swimming in smooth arcs about the surface, are another common sight. They send out a series of ripples, and by "reading" the tiny wave reflections with their sensitive antennae, they detect and home in on struggling insects.

Backswimmers suspend themselves upside down at an angle just below the surface. They dart up quickly to prey, using their long back legs as oars. Farther down, in the murkier depths, lurks the giant water bug, a formidable hunter of tadpoles and young salamanders.

79

SCALY VISITORS

Snakes and lizards belong in any garden that aspires to be a balanced community of wild creatures.

Reptiles do not usually number among the animals gardeners deliberately attract to their backyards. While turtles are one of the gardener's favorites, snakes and lizards suffer from a fearful reputation. Serpents, in particular, evoke a combination of phobia and fascination. Yet snakes can help the gardener control the proliferation of plant-eating creatures. As highly skilled predators, they hold a valid position towards the top of the natural food chain.

SNAKY FRIENDS

The snakes that commonly dwell in backyard gardens are benign towards people. Of these, the garter snakes are most prevalent. Active during the day, the common garter snake is either docile or quickly retreats if approached. Gardeners will appreciate its taste for slugs and snails. It also takes frogs and tadpoles.

Another handsome species that has prospered in proximity to people is the milk snake. Highly variable in appearance over its broad range, its markings always include some blotches of red bordered in black. Larger than a garter snake, adults can approach 6 feet (2 m) in length, and occur in a wide range of habits—woodlands, suburbs, pond margins, farmlands, and, especially, rocky sites.

Gardeners are more likely to encounter this nocturnal snake when overturning a log. It is an important predator of rats and mice, which it kills by constriction. Folklore wrongly alleges this snake has a malicious ability to steal milk from cows—hence, its common name.

THE COLLARED LIZARD

(top) eats insects of many kinds, including grasshoppers. The deadly poisonous copperhead snake's coloration (right) provides good camouflage from predators. The common garter snake (below) favors damp meadows, dense groundcover, and the margins of ponds and streams.

PERIL UNDER FOOT

Killing snakes is rarely justified, even in panic, since the wanton slaughter includes many benign, indeed useful, species misidentified as potentially lethal. And despite venomous snakes' reputations, they play an important part in the garden's ecology.

Be aware, though, that there *are* 19 fanged, venomous species in North America. Wearing high boots is a good idea when you are establishing a garden or tending cultivated edges that blend into their natural habitat. Avoid putting your hands or feet where you can't see, such as in crevices, under rock ledges, or in any possible hiding place for snakes.

LIZARDS AND SKINKS

Lizards are closely related to snakes and share a similar impervious, scaly skin; but with rare exception, they are four-legged. They are most populous and diverse in arid, rocky locations in the West and Southwest, but are by no means limited to hot deserts.

Lizards are also found in the more humid climes of the Southeast—the green anole, for example. Several other slender lizards, called skinks, also occur in the Southeast. Skinks are famed for their breakaway tails that thrash for several minutes on their own after being detached. This distracts an attacker and allows the skink to escape.

A PLACE FOR LIZARDS AND SNAKES

To encourage reptiles in or near your garden, and to enjoy a chance to observe them, provide a pro-tected, sunny, south-facing,

rock-laden hillside or wall, where radiant warmth will benefit their cold-blooded bodies. Leafy groundcovers and mulched, exposed soil will favor them by providing a cool retreat during the hottest sunshine. A reptile-friendly garden can include some easy-to-maintain features: a casual, stony bank, a rock pile, or a decaying woodpile.

INSIDE THEIR SHELLS

The other major group of reptiles is the turtles. Most turtle species make their homes near water, especially in ponds. Consequently, there are more turtle

THE GREEN ANOLE *(above) readily ventures into Southeastern suburban backyards, and is a useful consumer of plant-eating insects. Painted turtles (below) often congregate on half-submerged logs to bask in the sunshine.*

species in the wetter East. The painted turtle, found in ponds, streams, and lakes throughout the eastern and central United States, is one of the most common species. The reddish markings on the edges of its shell, along with red or yellow stripes on its head and feet, give it its name. Its diet is broad-ranging, from aquatic plants to tadpoles and snails. Along the West Coast, the western pond turtle fills a similar niche and lives in lushly vegetated ponds and streams.

HUMMINGBIRDS

*Tiny, delicate hummingbirds are the only birds
on Earth that have the ability to fly backward.*

H ummingbirds are
the smallest of birds,
delicately propor-
tioned, with very slender bills
and sweeping, narrow-tipped
wings. At the same time,
they are the fastest and most
energetic birds to visit the
garden, fierce in their defense
of territory and ethereal in
their courtship displays.
Capable of nearly instantaneous
changes in direction and speed,
hummingbirds hover in midair
with superlative ease. Their
ability to fly backward allows
them to back out of deep,
tubular flowers. This incom-
parable agility suits their
specialized role as nectar
feeders, for which they have
no serious rivals among the
other birds of North America.

ANNA'S HUMMINGBIRD *(top),
depicted here by John James Audubon,
is a resident of California, while the
ruby-throated hummingbird (right),
shown here at the nest, is the only
species found in the East.*

ENERGY AND SPEED IN A SMALL PACKAGE

As they gather the sugary
fuel, in the form of
nectar, to meet their
high energy demands,
hummingbirds astonish us
with their deliberate speed,
visiting many flowers in quick
succession, each for a fleeting
moment. In fact, they can't
keep up this breakneck pace
constantly, and like all animals,
they need to rest. So it's not
unusual to find them perching
quietly for five to 15 minutes;
during these pauses, energy
needs drop to a sixth of that
required when flying. While
taking these breaks, they
slowly digest nectar stored in
their small crop, a pouch-like
receptacle inside their throat.

*Everything about a
hummingbird is
a superlative.*

TOM COLAZO
(20th century), American naturalist.

At night, when they cannot
feed, their metabolic rate can
drop even further to conserve
energy. Their heart slows from
1,200 to 50 beats per minute
and their body temperature
cools as they enter a state of
torpor. This energy-saving
strategy is particularly useful
during spells of cold or in-
clement weather, such as
during the birds' spring and
fall migration periods, when
nectar-rich flowers are scarce.

RED FLOWERS *are particularly attractive to all hummingbirds, including the white-eared hummingbird (right).*

Body fat is important as fuel for long migrations. Before their travels between North America and their winter haunts in Central America, ruby-throated hummingbirds feed feverishly on flower nectar to add 50 percent to their weight in the form of fat. Such energy stores are altogether necessary if these tiny birds are to accomplish their 600-mile (1,000 km) crossing of the Gulf of Mexico.

HUMMINGBIRD FLOWERS

The flowers that provide hummingbirds with their energy receive the benefit of cross-pollination by these highly mobile feathered visitors. Some flowers specialize in attracting hummingbirds as pollinators, increasing the likelihood that their pollen will be carried to other plants of the same species.

Typical hummingbird flowers have deep, tubular corollas, requiring the long bills, protruding tongues, and artful hovering approach of these birds to reach their nectar-rich centers, and rendering them inaccessible to most flying insects, such as bees. Hummingbirds are drawn to red, in particular, but not exclusively. Other favored colors in approximate descending order of preference are orange, yellow, pink, and purple. Nearly any intense color will gain the attention of hummingbirds, and once they discover a recurrent source of nectar, they will remember it and will revisit flowers displaying that signal hue.

THE AERIAL FEATS OF A SMALL BIRD

The aerial feats of hummingbirds are carried out so smoothly as to appear effortless, while, in fact, these maneuvers cost highly in energy. Relative to their small size—many are no more than 3½ inches (9 cm) long—they consume nearly three times the calories of other garden birds on a busy summer day. The impressive flight muscles amount to about 30 percent of a hummingbird's body weight, which is proportionally larger than that of any other bird. This uniquely well-developed musculature provides a powerful upstroke as well as the customary downstroke that powers the flight of other birds.

The key to the aerial versatility of hummingbirds lies in their ability to rotate their powerful wings rapidly and in a number of different ways at the shoulder joints, an unparalleled innovation in the engineering of bird flight. The illustrations on the right show the three basic flight modes. Normal forward flight (Figs 1 and 2) is made possible by the up-and-down wing strokes that provide forward momentum and lift, at the rate of 80 beats per second. To hover in midair (Fig. 3), the wing tips move through a horizontal figure-of-eight at the rate of 55 beats per second. To back away from a flower (Fig. 4), the wings rotate in a circle above and to the rear of the bird at 61 beats per second, providing some reverse momentum in addition to lift.

The tail serves as a rudder, enabling sudden stops and changes in direction, and the audible whir of their wings produces the humming for which these birds are named. Males execute spectacular, deep dives and U-shaped swoops, which they often finish by hovering and flashing the jewel-like throat patch (gorget) in front of a prospective mate.

❶

❷

❸

❹

83

PLANTS TO ATTRACT HUMMINGBIRDS

To encourage long-term residency by hummingbirds, design your garden to include suitable nectar-bearing plants that flower through the different seasons.

Common name	Scientific name	Type	Flowering time
glossy abelia	*Abelia* × *grandiflora*	shrub	summer–fall
California buckeye	*Aesculus californica*	tree	spring
red buckeye	*Aesculus pavia*	shrub	spring
mosquito plant	*Agastache cana*	perennial	summer–fall
century plant	*Agave* spp.	shrubs	summer
columbine	*Aquilegia* spp.	perennials	spring–summer
trumpet creeper	*Campsis radicans*	vine	summer
flowering quince	*Chaenomeles japonica*	shrub	spring
desert willow	*Chilopsis linearis*	tree	summer
ocotillo	*Fouquieria splendens*	shrub	spring
fuchsia	*Fuchsia magellanica* & hybrids	shrubs	summer–fall
daylilies	*Hemerocallis* hybrids	perennials	summer
red yucca	*Hesperaloe parviflora*	shrub	spring–fall
coralbells	*Heuchera sanguinea*	perennial	summer
desert lavender	*Hyptis emoryi*	shrub	spring
scarlet gilia	*Ipomopsis aggregata*	biennial	summer
irises	*Iris* spp. & cvrs	bulbs, perennials	spring–summer
chuparosa	*Justicia californica*	shrub	fall–winter
lilies	*Lilium* spp.	bulbs	summer
cardinal flower	*Lobelia cardinalis*	perennial	summer
honeysuckles	*Lonicera* spp.	vines, shrubs	spring–summer
lupines	*Lupinus* spp.	annuals, perennials	spring–summer
bluebells	*Mertensia virginica*	perennial	spring
scarlet monkeyflower	*Mimulus cardinalis*	perennial	summer
four o'clocks	*Mirabilis* spp.	perennials	summer
bee balm, bergamot	*Monarda* spp.	perennials	summer–fall
flowering tobacco	*Nicotiana alata*	annual	summer–fall
garden geraniums	*Pelargonium* spp. & cvrs	perennials, shrubs	summer
penstemons	*Penstemon* spp.	perennials	summer
summer phlox	*Phlox paniculata*	perennial	summer
flowering currants	*Ribes* spp.	shrubs	spring
sages	*Salvia* spp.	perennials, shrubs	spring–fall
figworts	*Scrophularia* spp.	perennials	summer
scarlet hedge nettle	*Stachys coccinea*	perennial	summer–fall
cape honeysuckle	*Tecomaria capensis*	vine, shrub	fall–winter
California fuchsia	*Zauschneria* spp.	perennials	fall–winter

HUMMINGBIRD FEATHERS

Hummingbirds dazzle the eye not only with the deftness of their flight but also with the shimmering beauty of their plumage. So gem-like are the feathers of the male birds that Europeans used their plumage to make jewelry, and Mesoamerican cultures used it in ritual adornment. Shown here, from left to right, are feathers from Anna's, blue-throated, Costa's, and black-chinned hummingbirds.

THE COSTA'S *hummingbird (left) breeds in the Southwest. The pan-shaped feeder (above) features an ant moat at the top that prevents ants reaching the feeder's sugar solution.*

SUGARY SUPPLEMENTS

Gardeners may attract hummingbirds more reliably and in greater numbers by maintaining one or more feeders filled with sugar water. Feeders complement flowers by providing a dependable source of energy when natural nectar becomes scarce, as may happen during unseasonable cool, cloudy weather or simply during lulls in garden blooming. The recommended recipe for this nectar substitute is a maximum of one part sugar to four parts distilled water, boiled briefly to kill off any bacteria. More concentrated solutions may cause liver damage in hummingbirds over the long term. Anything weaker than a one to eight solution (11 percent) is unlikely to satisfy their energy needs. Red dye in the water is unnecessary, though something brightly colored, such as a red ribbon or the red plastic collars that surround the tubes on some feeders, will help to grab the attention of hummingbirds in your neighborhood. Never use honey because it sometimes harbors a fungus that is deadly to hummingbirds.

Replace the feeder solution and clean the sugar water receptacle and tubes on a regular basis—at least once a week is recommended. Any black mold that appears can be cleaned out using vinegar; avoid using soaps or detergents, which could leave a harmful residue.

Other birds, especially orioles and house finches, will occasionally visit hummingbird feeders with perches.

THE BROAD-TAILED HUMMINGBIRD *(left) is a resident of the Rockies, while the rufous hummingbird (right), depicted here by John James Audubon, can be found in the Pacific Northwest.*

Wasps and bees are also drawn to sugar–water feeders. When their number rises to the point where they intimidate the hummingbirds, cover the end of the feeding tube with a plastic mesh protector that prevents bees from directly sipping the solution; placing the feeder in shade will also deter these and other insects. Hummingbirds don't mind shade and in fact, seek out a cool perch on occasion.

INSECT-EATING BIRDS

Many bird species readily adapt to our gardens
and hold the greatest immediate appeal for gardeners.

With their mantle of feathers, birds share the human emphasis on colorful visual display, and their song is often melodious to human ears. They are an interesting spectacle: not too small and given to a warm-blooded liveliness in their feeding and social behavior. Indeed, they not infrequently raise their young right before our eyes. Many bird species will quite willingly gravitate to our yards for food when we offer it. Furthermore, it is exceedingly easy to design planting schemes that satisfy both their needs and our aesthetic goals.

An attentive gardener could easily record a dozen species of bird in a garden designed for habitat diversity. A core group will comprise year-round garden residents.

INSECTS AND MIXED DIETS

To attract the greatest variety of birdlife, you must take into account the critical insect component of the avian diet. Chicks need protein in their diet to develop from fragile hatchlings to robust flyers, an accelerated process that takes only a matter of weeks. So most songbirds raise their young during the flush months of summer, when insect populations, the nutritional answer to their needs, are surging. This is the time of year that the gardener is most likely to encounter birds as consumers of insects and other invertebrates. In the other seasons, fruit and seeds are likely to figure more significantly in bird diets. Even in winter, however, some species still focus primarily on insects (especially eggs and pupae).

PLANTS THAT ARE HOST TO INSECTS

Many plants that develop seeds and fruits in summer and fall host a rich array of insects when the plants are blooming and leafing out in spring. Densely vegetated meadows and borders composed of herbaceous plants are a rich, buggy environment during their bloom season, visited by many insect-eating birds. Flowers of all kinds attract a host of six-legged pollinators and pollen eaters, which become the prey of birds such as flycatchers, bluebirds, and others. The supple, young foliage of some common native trees, such as oaks and willows, is initially free of insect-repelling tannins, and, along with the flowering catkins, support an abundance of spring caterpillars. Large numbers of migrating song-birds, particularly warblers, vireos, and tanagers, feed on this springtime bounty as they pause on the way north to their breeding grounds. The bark of trees of all sizes and types is home to insects, spiders, and pupae, which birds such as nuthatches, chickadees, and the brown creeper frequently seek out while a part of mixed feeding flocks.

THE TUFTED TITMOUSE *(top) uses its small, short bill to pick off tiny insects from branches and bark. This female indigo bunting (left) is feeding her chicks the protein they need in the form of juicy, green caterpillars.*

THE SCARLET TANAGER *(above)
has a slightly hooked bill to snare large
prey, while the pileated woodpecker,
(right), recognizable by its red crest,
excavates deeply in tree bark for insects.*

CATCHING INSECTS

Insect–eating birds have strong, yet nimble feet that are variously adapted to perching on the slenderest of twigs, hanging upside down, walking up trees, or hopping across the ground. The bills of insect-eating birds are suited to seizing crawling and flying insects, picking away at tiny bugs and cocoons, pounding into trees, or crushing hard-shelled beetles and other specialized tasks.

The most common method of insect-eating birds is to glean plant-eating insects directly off foliage, stems, and flowers. Warblers, vireos, and tanagers adopt this approach, with various species focusing on different levels and types of vegetation. Using their slender bills, warblers hunt leaf-loving creatures at a speedy pace, hovering and flitting in pursuit of caterpillars and other small prey items. By comparison, vireos and tanagers have marginally heavier bills, slightly hooked to snare larger prey,

which they stalk in a more deliberate fashion. Catching flying insects on the wing is the preferred method of the agile flycatchers, whose bills are more gaping and whose long, narrow wings enable deft midair turns.

Woodpeckers excavate for prey in decaying trees, using their sharp, powerful bills and extensile tongues, with their feet firmly anchored to the trunk or branch. They sometimes plumb the ground for ants and visit feeders.

Yet other species operate at ground level, walking, hopping, or scuffling

BLACK-CAPPED *chickadees
(far right) nest in tree cavities
and will accept nestboxes.
Bullock's orioles (right) feed on
insects and fruit alike during their
spring and fall migrations.*

while they search through leaf litter for crawling insects and other invertebrates. The excellent vision of song-birds serves not only to detect elusive prey, but also to keep a watchful eye out for predators, whether they be gliding hawks or terrestrial mammals.

87

BIRDS *that* EAT FRUIT *and* SEED

The fruit and seed of many plants are often deliberately conspicuous and alluring to birds.

When birds pursue prey such as insects and other invertebrates, they frequently must contend with elusive, moving targets and concealed quarry. Fruit and seed are much more easy prey. Plants trade the transport of their viable seed to new locations in exchange for providing food (carbohydrate, fat, protein, and vitamins) to birds, as well as to some accommodating mammals— a mutually satisfactory arrangement. In this way, fruit- and seed-

SOME MEMBERS *of the pheasant family, such as the Gambel's quail (below) of the Southwest, visit gardens in search of fruit and seed. Typical of most sparrows, the dark-eyed junco (below right) uses its stout bill to crush seed and fruit, and sometimes, insects.*

eating birds play a direct role both in the dispersal of seed far from parent plants and in the spread of many plant species to new locations and environments.

WAYS PLANTS CAN DISPERSE SEED

Plants have evolved various strategies to accomplish this vital goal of seed dispersal. Brightly colored fruits with agreeably pulpy, nutritious flesh initially attract birds. After the birds have had their fill, they frequently discard the pits if they are large and hard to crack open. The bird sometimes drops the pits only after it has flown to a new perch, transferring the seed within to a new location. The smaller seed of some fruiting plants, such as the barberries (*Berberis* spp.), will withstand bird digestion intact and is transported for as long as that process takes. The successful fruit trees belonging to the *Prunus* genus—apricots and cherries—have bitter, even poisonous seed that discourages the bird from eating after it has consumed the tasty flesh of the fruit.

Birds, the free tenants of

land, air, and ocean,

Their forms all symmetry,

their motions grace.

JAMES MONTGOMERY
(1771–1864), British poet

Employing another clever strategy, the seed inside the fruit of the mistletoes is so disagreeably sticky that birds, after sating their appetites, often rub the seed off their bills onto tree bark, exactly where this parasitic plant prefers to grow.

Many species of plant do not coat their seed in an inviting rich fruit wrapping, but suffer some of it to be broken open and eaten, relying on the minority that is scattered or misplaced by foraging seed eaters, such as sparrows and squirrels, to perpetuate their line. The seed heads of certain plants, such as *Impatiens*, burst open when birds start to peck them, propelling some seed a few feet away. In other plants, the seed simply shakes free when the birds alight, so that some escape being eaten. In watery environments, seed blows down to rest at the muddy water's edge and then travels

THE PURPLE FINCH *(right) eats primarily seeds from grasses and herbaceous plants. The seed-eating cardinal is a year-round resident in the eastern and central United States; the male (far right) is bright red.*

in the mud picked up on the feet of marsh and pond birds. Still other seed, such as that of the tickseeds (*Bidens* spp.), has tiny hooks that allow it to attach to feathers and enjoy a considerable ride away from the parent plant.

FICKLE VISITORS AND FLEXIBLE PALATES

Be prepared for inconsistency in the food preferences and appetites of birds. They will patronize some fruit-bearing plants immediately, yet ignore other fruits until they season naturally and become less astringent, or until other preferred food resources have been exhausted. While one year a fruiting shrub may attract a resident flock of waxwings; another year, it will fail to draw much at all in the way of feathered visitors (though always providing a welcome touch of color for the garden). In fact, birds are not only fickle but they also are surprisingly flexible, and when times are lean, they will accept food that they would otherwise ignore. For the early-spring migrants, such as phoebes and bluebirds, which sometimes get caught in a cold snap when most insects are dormant, the lingering berries in the yard will prove highly attractive as a vital, emergency food source. Even in the summer, when insects are abundant, some primarily insect-eating birds will then indulge themselves in berry-eating to take advantage of certain minerals and vitamins.

GARDEN SEEDS AND FRUITS FOR BIRDS

A meadow planted with native grasses and herbaceous plants—such as sunflowers (*Helianthus* spp.), wild iris (*Iris* spp.), asters (*Aster* spp.), goldenrods (*Solidago* spp.), *Coreopsis* spp., and buckwheats (*Eriogonum* spp.)—will provide the seeds that attract sparrows, juncos, buntings, and finches. Some seed heads last well into winter to serve as a source of protein and fat for birds.

Introducing berry-bearing shrubs and small trees is another productive strategy for luring a variety of birds, particularly in fall and winter. Possibilities include winterberry (*Ilex verticillata*), in the East, and toyon (*Heteromeles arbutifolia*), in the West (see The Backyard Habitat, starting on p. 100, for other species suitable for your region). There are also some exotic, non-invasive plants worth considering. Typical bird species drawn to these plants include thrushes (such as the American robin), waxwings, jays, thrashers (and the related mimic thrushes such as the catbird), bluebirds, and occasionally, warblers and tanagers.

THIS CEDAR WAXWING *(below) is feeding its chicks pin cherries (Prunus pensylvanica).*

Small Mammals
in the Garden

Many small mammals benefit the natural garden by preying upon a variety of plant-eating insects.

Appreciating the mammals that regularly dwell in the garden presents a challenge. On the one hand, these furry, warm-blooded animals are closer to us on the evolutionary scale than any other garden creatures. On the other hand, their habits make them difficult to observe—they tend to be secretive. To rest and breed safe from predators, they often burrow underground or hole up out of sight in trees. Many species' peak activity above ground occurs only under cover of night.

We must often rely on indirect evidence to detect mammals—the tell-tale traces of their activity, such as footprints, trails and runways, and gnaw-marks. Their calling cards sometimes include shed hair, droppings, chewed stems, and other remains of a vegetarian meal. Yet, those same resourceful and energetic natures that at times bedevil us can also amuse us and arouse our respect.

THE SOUTHERN FLYING SQUIRREL
(top) does not actually fly, but glides via a cape of skin between its front and hind legs. It roosts inside tree trunks. The white-footed mouse (right) is found across most of the United States, except the Southeast and the West.

THE MASKED MARAUDER
No mammal better exemplifies this paradox than the raccoon. Though technically a carnivore, its diet includes fruits and nuts, plant-eating insects, birds' eggs, carrion, and occasionally, small, live animals such as mice and rats.

Suburban yards and even the congested urban landscape are home to raccoons. In their search for food, raccoons are not in the least averse to pawing through our garbage. They also use their nimble paws to pick grapes and figs, and they are able to scoop out the flesh of a watermelon after making just one small hole.

GROUND-LEVEL HUNTERS
The striped skunk is another mammal that has thrived in proximity to human habitation across most of the continent. It is notorious for

its pungent discharge, and its bold black-and-white coloration probably evolved to warn predators of this powerful deterrent. In the garden, the striped skunk preys upon insect grubs, and eats a large measure of berries in summer and fall.

THE CHAMPION INSECTIVORES
Shrews' high-strung ferocity has few parallels among mammals. They dart through leaf litter, furiously attacking such prey as insects, centipedes, and snails with their sharp teeth, and consuming up to twice their own body weight each day.

Moles, though related to shrews, aren't so hyperactive, but as master underground diggers, they are very efficient hunters nonetheless. Their powerful, clawed feet and svelte, compact bodies allow

THE EASTERN CHIPMUNK *(left)*
stuffs nuts into its ample cheek pouches
before taking them back to its burrow.
The marsupial opossum (above) nurses
its young inside a special pouch.

them to virtually swim through soft soil in pursuit of earthworms and insects. While moles may damage lawns and plant roots, they also benefit the garden by aerating soil and reducing populations of root-eating grubs.

THE PLANT NIBBLERS

Mice and voles are true rodents, and feed on plant material, particularly grasses, seeds, nuts, and fruits, as well as mushrooms and small insects. The deer mouse is the most common species across much of North America.

Rabbits use their prominent front teeth to browse grasses and other herbaceous plants, as well as buds, twigs, bark, and leaves of woody plants. Eight species of rabbit are native to North America, most of them ranging south of Canada. The snowshoe hare is widespread across Canada.

THE NUT GATHERERS

The eastern gray squirrel is surely the most conspicuous and acrobatic rodent to visit gardens in the eastern half of the continent (as far north as southern Quebec). Its two cardinal characteristics are its antic tree-dwelling behavior and its extravagantly bushy tail.

The agility of the squirrel serves it well in jumping and running through its arboreal domain, gaining it ready access to nuts, buds, and blossoms, as well as escape from predators.

Gray squirrels primarily eat acorns, hickory nuts, and beechnuts. With the approach of winter, eastern gray squirrels become inveterate hoarders, compelled to gather and bury a multitude of nuts. Since many nuts go unrecovered,

RACCOONS *(right) prefer hollow trees*
or rocky crevices in which to rest and
to raise their young.

squirrels play a significant role in propagating the very trees that provide the mainstay of their diet. Gray squirrels also relish the berries of such ornamental shrubs as holly (*Ilex* spp.) and *Cotoneaster* spp.

In the West, nut-gathering, tree-dwelling squirrels are the western gray squirrel, ranging from California to central Washington, and the tassel-eared Abert's squirrel, found in some of the mountains of the Southwest. Neither one enjoys the same reputation as its eastern relative for boldly cavorting about backyards. The California ground squirrel sometimes ventures into garden meadows and farm fields.

LARGE MAMMALS
in the LANDSCAPE

The larger mammals have a special hold upon the human imagination.

Dramatic and hand-some in appearance, fast and muscular in their physical abilities, and wily in their habits, large mammals arouse our fascination and admiration. A greater degree of mystery adheres to these widely roaming animals than to their smaller cousins, and their occasional appearances are thus more treasured.

With the exception of deer, the coyote, and the red fox, populations of the larger mammals continue to decline in the face of human en-croachment, to the point where most species are rarely sighted anywhere close to civilization. Ironically, their near–disappearance, which we caused, now spawns a desire to preserve vestiges of the wilderness they require. With greater weight given to their habitat requirements, our chances to witness these most dramatic of native mammals are sure to be enhanced.

HOOFED HERBIVORES
Of the larger animals, the two common deer species, the white–tailed deer and the mule deer, have adapted well to human civilization. The white–tailed deer ranges throughout most of North America, except California and the Southwest. The closely related mule deer, which gets its name from its large, expressive ears, is common throughout most of western North America.

Male and female deer (bucks and does) tend to gather in separate herds to wander wooded and open landscape in search of foliage. Well-beaten paths through grass and shrubbery are a sign of their presence. They are most active around dawn and dusk.

Other, more indirect signs of deer in the local area relate to their diet. In winter and spring, they browse on woody plants from

IN LATE SUMMER, *bucks such as this mule deer (below) will rub the velvet on their antlers off on tree trunks. Deer browse on woody plants from the ground level up in winter. This white-tailed doe and yearling (bottom) are availing themselves of a tasty shrub.*

ground level to as far as they can reach. During winter, especially where snow falls heavily, they turn increasingly to the foliage of evergreens, such as hemlock and even the yews that are toxic to other animals. When the first green deciduous leaves appear, their diet shifts to herbaceous plants such as grass, wildflowers, and the succulent leaves of certain shrubs, vines, and trees. Many deer avail themselves of our own lush, summer plants and the bark of some ornamental trees and shrubs, and often do so quite boldly in suburban settings where hunting is not common. Whether in a wild or settled area, acorns are a favorite for deer in fall.

DEER IN THE GARDEN

To attract these lithe, statuesque animals, provide apples or fleshy vegetables at low feeding stations, or simply tolerate their presence in an orchard or informal squash patch. Salt blocks for licking will also draw deer: all animals need salt in their diet, and will take it when it is offered. In fairness to the well-being of nearby gardens, you should consult with neighbors before encouraging visits by deer. Planting dense stands of evergreens, especially if interspersed with low shrubbery or meadow, will provide deer with the cover they seek for their periods of rest. Where winters are snowy and severe, deer herds gravitate to evergreen thickets. Winter is the time of greatest stress for deer, and, quite sadly, with the decline in their natural predators, deer populations in many parts of the continent rise out of control until slow starvation takes its toll.

THE CANINE CARNIVORES

In earlier centuries, the gray wolf, hunting in packs—as well as the mountain lion—helped keep deer populations in check. Though the gray wolf has disappeared from nearly all its range south of Canada, its close relative, the coyote, still manages to survive, even thrive, across the continent, despite vigorous campaigns to hunt and poison it out of existence. It preys adeptly on virtually any animal it comes across, including large insects, songbirds, amphibians, rodents, and deer.

ONLY THE LUCKIEST GARDENER *will ever catch sight of the bobcat (top). Although common across much of North America, it is skilled at hiding in concealing shrubbery. A taste for corn, berries, and fruits draws the red fox (above) into our gardens. These pups will also learn to kill the live prey their parents bring them.*

Extremely wary of humans, coyotes generally stay out of sight, though they probably stray into gardens surreptitiously in rural areas.

The gardener in rural and suburban areas has a better chance of encountering the red fox, another clever hunter, though like other members of the canine family, it is skilled at detecting and avoiding humans, relying on its acute senses of smell and hearing. This generally quiet, quick-footed species prefers forest edges and meadows, where it pounces upon voles, rabbits, crickets, and other insects, most often at night and in the early morning.

BEARS HAVE BEEN *known to visit garden berry patches at the forest edge in remote areas. Be aware of the danger.*

NIGHT DWELLERS

*When the sun sets, darkness envelopes the garden
and a new cast of wildlife characters stirs.*

Spiders freshen their webs; nocturnal animals scurry or take wing; tree frogs trill; fireflies dazzle at the meadow's edge—a different world emerges at nighttime.

ENTERING THE NIGHT WORLD

The after hours world is by no means closed to the curious gardener who is willing to adapt. Learn to rely on non-visual senses to appreciate the tell-tale sounds and fragrances that rise in importance at nighttime. And take advantage of what light is available. For example, the twilight provides an opportunity to glimpse night dwellers as they renew their activity following daytime

slumber. Against the dimming sky, bats and owls take flight in silhouette. Some animals, such as the Pandora sphinx moth, are at their peak activity during this time of transition. A garden bathed in moonlight is similarly open to our perusal.

Even in starlight, our eyes can adjust to see shape and movement in the garden. The human eye achieves surprising nighttime sensitivity after only 15 to 30 minutes in darkness away from artificial lights.

NOCTURNAL FLOWERS AND THEIR ADMIRERS

Like us, most animals can't see in color at night, so it's no surprise that pale-colored flowers, especially white ones, stand out in the night garden. Some flowers glow to their full glory only at night, such as the moonflower (*Ipomoea alba*) with blooms that last a single night. Other flowers advertise nectar and pollen with bold, often exotic fragrances released only after the sun has set. Many of the lilies, flowering tobacco (*Nicotiana alata*), and, in mild-climate gardens, angel's-trumpets (*Brugmansia* spp.) and night jessamine (*Cestrum nocturnum*) are prime examples of nocturnal fragrance. The increased humidity of night air enhances such fragrances,

which communicate with potential pollinators.

Night flowers usually recess their nectar inside long, tubular blossoms, accessible to the long tongues, or probosces, of the night-flying sphinx moths (also known as hawk moths). Patient nighttime observers may spot these moths hovering like small, fluttering ghosts at flower borders.

Moths do not enjoy a monopoly on nocturnal nectar sources: for example, in the Desert Southwest, the famed saguaro cactus is pollinated by the long-tongued bat as it laps up springtime nectar.

THE LOVELY LUNA MOTH (top), *a nocturnal wanderer, is an endangered species. The pyralis firefly (above right) has a yellow flashing light on its tail. Its larvae feed on slugs and snails. True to its name, the barn owl (left) often nests in barns and silos.*

THE CALIFORNIA LEAF-NOSED BAT
(above) preys on various insects, including flightless ones. Some spiders, such as this green lynx spider (right), use a silk dragline when pouncing on prey.

SWEEPING THE SKY

Other animals, principally bats, naturally take advantage of night–flying insects as a food resource. The key to the bats' success is echolocation, which they employ to zero in on prey and to navigate through the landscape. Bats emit high-frequency sound waves in flight that bounce off objects ahead of them and then rebound to their perceptive ears for interpretation. They can track an insect as small as a mosquito.

Perhaps because they are active only at night, when most of us are asleep, or because they have been the featured villains in so many horror movies, bats have been saddled with a negative reputation, which does not reflect their real value in the ecology of many regions around the world. Throughout North America, small bats such as the little brown bat play a major role in the reduction of night-flying insects, particularly mosquitoes. In the tropics and desert regions, bats play a crucial role in the lives of many plants native to these

Sit outside at midnight

and close your eyes;

feel the grass,

the air, the space.

LINDA HASSELSTROM (20th century), American writer and rancher

areas, either by pollinating flowers in their search for nectar, or by dispersing seed as they consume the fleshy fruit of cacti and tropical trees.

HAWKS OF THE NIGHT

Alongside bats, but in far fewer numbers, birds called nightjars sweep the insects out of the night air into their wide, gaping mouths. In one evening, a nightjar can capture 500 mosquitoes. Soft-edged flight feathers muffle their aerial arcs

above woodland edge and garden. From fence post or branch, they sing the rich, melodious, and oft-repeated songs for which many of them are named, such as "Chuck-wills-widow", or "Whip-poor-will".

Sharing the nightjars' capacity for agile, silent flight and night-piercing calls, owls cruise the nocturnal landscape in search of larger prey. For example, the small, mottled screech owl is adept at capturing mice and voles, as well as grasshoppers and cutworms. It nests in tree cavities and prefers gardens with groves or orchards nearby. It doesn't actually screech, but utters an eerie whistle. The larger barn owl hunts over fields, farm-land, and gardens, descending on as many as 20 rodents in a single night.

WILDLIFE
MISCHIEF MANAGEMENT

To enjoy the privilege of having wildlife in the garden, gardeners can use ingenious, nonlethal measures to limit problem animals.

Under the best of circumstances, the natural garden is a balanced and diverse wildlife community that is essentially self-regulating. However, certain mammals, in particular, are so voracious that they are not easily tolerated.

FENCES MAKE GOOD NEIGHBORS

The most effective strategy is to use a fence or other barrier to prevent nuisance animals from entering the garden in the first place. The jumping, climbing, squeezing, and burrowing abilities of the animal(s) will dictate dimensions: height off the ground, depth below ground, and size of openings.

Excluding deer presents the tallest challenge because of their jumping ability. While 7-foot (2.1 m) high fencing usually suffices, 8 feet (2.4 m) is thought to be more deer-proof. Wire mesh, high-tension or hog-wire fencing, and wood fences are all good options. A doe can slip through an opening as tight as 12 inches (30 cm), and fawns need only 10 inches (25 cm). Deer will first try to pass under a fence, so secure the base at ground level. Check the fence regularly for breaks and weaknesses.

Raccoons cannot jump but their climbing ability renders them a difficult animal to exclude. Wire-mesh fencing must include a wire-mesh roof to be effective. Or leave the final 2 feet (60 cm) of a 6-foot (1.8 m) fence loose and floppy so that raccoons cannot climb it, and fall backward to the ground. Insert the fence at least 6 inches (15 cm) into the

INTERPLANTING *nasturtium (Tropaeolum spp.) with more vulnerable plants may deter woodchucks (above) from the garden.*

FENCES FOR BURROWING ANIMALS

Fences must extend into the soil to stop burrowers entering the garden. An 8-inch (20 cm) flange at the bottom turned outward is also recommended to thwart burrowers excavating down and under the barrier.

Never use barbed wire, because it could injure wildlife, as well as pets and errant children.

Animal	Size of mesh	Depth in ground
Gophers	½ inch (12 mm)	24 inches (60 cm)
Moles	½ inch (12 mm)	18 inches (45 cm)
Rabbits	1 inch (2.5 cm)	12 inches (30 cm)
Woodchucks	1 inch (2.5 cm)	18 inches (45 cm)

THE OPPORTUNISTIC RACCOON
(far left) likes a broad range of food, including tidbits left for the birds, while cats (left) can seriously deplete native bird populations. This eastern cottontail (below) is surveying its next meal.

naturally pungent qualities and from carnivorous animals. Consult your local organic garden or ecology center for safe recommendations suited to your region.

Electric devices include those that produce harsh sounds to unnerve animals, and gadgets that reputedly create magnetic fields to repel some mammals. Be forewarned, though, that product claims may exaggerate their effectiveness.

PLANTING STRATEGIES

Interspersing vulnerable plantings with plants endowed with pungent foliage can be a successful strategy. The natural odor overwhelms the very sensitive sense of smell of many mammals, interfering with their enjoyment of nearby plants and sometimes actually making them turn away in disgust. The following plants are worth trying: wormwood (*Artemisia absinthium*); onions (*Allium* spp.); dusty miller (various species) and marigolds (*Tagetes* spp.), particularly for rabbits; caper spurge, or gopher plant (*Euphorbia lathyris*) and daffodil bulbs, for gophers and moles; squill bulbs (*Scilla* spp.), for gophers; foxglove (*Digitalis* spp.), for deer; *Salvia* and *Mentha* spp.

A more subtle method is to surround vulnerable plants with dense plantings of species that hold no dietary interest for browsing mammals. The point can be made more emphatically by creating an informal barrier from a thorny plant such as barberry (*Berberis* spp.).

soil to prevent raccoons pulling up the bottom and squeezing through.

CAGES FOR PLANTS

Wire-mesh caging surrounding plants and tree trunks will provide selective shielding from hungry jaws. For deer, caging should extend at least 5 feet (1.5 m) high. Voles can be deterred from tempting tree bark by a band of hardware cloth 24 inches (60 cm) high.

To protect them from moles, plant bulbs and other perennial plants inside wire-mesh baskets or inside cans with holes for drainage.

Birds and raccoons can be deterred from grapes and other fruit either by covering the fruit in plastic netting once ripening starts, or by wrapping bunches of grapes and ears of corn in nylon stockings.

REPELLENTS

Next to physical barriers, repellents rank as the most useful remedies for problem animals in the garden. The most commonly used are derived from plants with

OBSERVING *and* RECORDING

Sightseeing opportunities abound in the natural garden planned for a diversity of wildife.

Witnessing the varied conduct of animals' lives—from small-scale events inside the corolla of a flower, to the dramatic migrations of mammals and birds—enriches the enjoyment of the garden beyond the satisfaction provided by the plants themselves. Sometimes, our experience depends on the interpretation of animal traces in the garden: a nibbled leaf, the tracks in freshly fallen snow, or the ant hill that indicates an elaborate network of tunnels under foot.

MOUSE TRACKS IN THE SNOW *(above) betray the presence of a furry visitor. Close observation reveals a well-camouflaged goldenrod spider (right).*

explain connections between different animals and plants and to record intriguing details. Cast away any concerns about artistic skill or scientific precision: an easy flowing pen or soft pencil, guided by a spontaneous hand and an alert eye, can complement your written observations and enhance subsequent recollection and understanding.

A GARDEN JOURNAL
A garden journal not only fixes memories of solitary events, but allows comparisons to be made and patterns to be discerned over time. Sketches are an invaluable supplement to this personal chronicle to

ORGANIZING THE DATA
To begin, record basic data such as temperature, rainfall, and date and time of day. It is also useful to categorize behaviors: for example, feeding, incubating, perching, calling.

Later on, you could organize wildlife observations by animal species or host plant. Index cards or a home computer equipped with a basic database program can organize data very efficiently.

THE EYE OF THE CAMERA
Photographs are valuable as a supplement to illustrate the history of a garden and its inhabitants and visitors. The 35 mm single-lens reflex (SLR) camera offers the best value for amateur nature photography. It allows the photographer considerable control over the image at a moderate price.

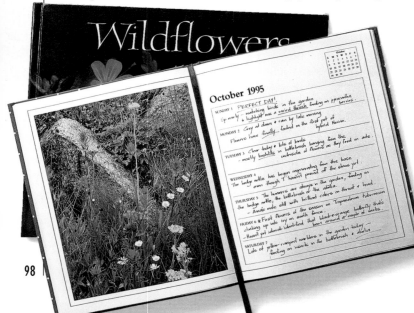

A GARDEN JOURNAL *Use your journal (left) to make connections between plants and wildlife. For example, a late year for the blooming of columbine might mean fewer springtime hummingbirds in the garden.*

You can use a variety of interchangeable lenses to suit different situations. A tripod for the camera in difficult, low-light situations will eliminate the blurring effect of accidental hand-held camera movement. Filters are also useful to compensate for color imbalance and to reduce glare.

TIPS FOR THE WILDLIFE OBSERVER

There are a few simple techniques to hone observation skills and improve the odds for witnessing wildlife behavior. Strive to reawaken your child-like curiosity and emulate the calm, deliberate focus of a detective.

Keep in mind how intimidating humans are to other vertebrate animals, especially birds and mammals. Strike a low, quiet profile: wear soft shoes and nonrustling clothes; avoid loud conversation and refrain from staring at animals that seem nervous. Above all, move slowly, pause often, and don't rush toward the focus of your interest.

Take a few minutes to listen carefully and deliberately to all the sounds, calls, and songs in the garden. Detecting the presence of cleverly camouflaged animals, such as katydids, may depend almost entirely on aural signs. Inhale the odors of the natural world: not just the obvious flower fragrances, but more subtle scents such as the spiciness of fall leaves, or the robust, mineral aroma of freshly turned earth. Use your hands to experience plant textures, from the downy underside of an oak leaf, to the elaborate ridges of fir tree bark.

DRAWINGS CAN ENHANCE *your garden journal, whether quick sketches or skilled renderings, like this watercolor of a blanket flower (Gaillardia aristata).*

Take in the whole of your garden's vertical range. Look down at ground level and under stones and logs for hidden inhabitants. Look up into the trees and overhead for soaring birds.

Schedule your observations to coincide with the peak activity of different animals. Some mammals prefer the dimly lit hours before sunrise and after sunset; others reach their daily zenith only in the middle of the night. Bird activity peaks after the first rays of sun have warmed up insects.

Take note of the first appearance of an animal: for example, the arrival of a colorful bird from its winter habitat in the tropics. Mark the beginning of a significant phase in a plant's yearly cycle.

INSTRUMENTS TO HELP YOU

• A rain gauge and minimum–maximum thermometer allow you to record and track the two most important elements affecting the garden and its wildlife, and help you to become more aware of the progression of the seasons.

• Collect small specimens for later identification and possible preservation or, in the case of leaves and flowers, pressing.

• Binoculars allow a much greater intimacy with shy and retiring backyard residents. The two most important technical criteria are the power of magnification, which should be at least 7 and no more than 10;

and the objective lens, a measure of light-gathering ability, which should be at least 30 and can be as high as 50. Binoculars sold in stores are usually 7 x 35, and these should suffice in any garden situation; 8 x 42 function more effectively in low-light conditions of dawn and dusk, when many mammals are particularly active.

• To reveal more detail about very small creatures and their plant environment, carry a simple magnifying glass with a folding handle. More costly pocket microscopes with battery-powered lights permit variable levels of magnification.

ARTS AND CRAFTS *have always drawn inspiration from nature, whether from the world of animals, or from plants and flowers. This Baltimore bride quilt (right), sewn in the 1840s, uses floral motifs in its design.*

Everybody needs beauty as well as bread,
places to play in and pray in, where Nature may heal
and cheer and give strength to body and soul alike.

JOHN MUIR (1838–1914),
Scottish-born American naturalist and writer

CHAPTER FOUR

The BACKYARD HABITAT

THE SIX REGIONS

Gardening climates and distribution patterns of natural vegetation have determined our regional divisions.

The 144 native plants in The Backyard Habitat have been grouped into six regions within the United States and Canada. These regions have been established based upon distribution patterns of the natural vegetation, combined with the gardening climates that distinguish various parts of the continent. Thus, for instance, the long, relatively rainless summers of the West Coast are distinct from the higher elevation arid regions of the Mountains and Basins.

Each region opens with a double-page artist's impression of a habitat garden showing typical plants and the visitors they are likely to attract.

Concluding each regional group of 24 plants is a list of an additional 20 plants that are good choices for a regional habitat garden; included in this list will be a few plants that are not native to the region, but are well-adapted, well-behaved (non-invasive), and are attractive to wildlife.

A small version of the regional map is reproduced for each species featured in The Backyard Habitat. The map shows in green those regions in which the plant can be found growing natively. Any other regions in which it might be expected to be easily cultivated are in orange. Be aware that each species is likely to have a natural range much smaller than the entire region or regions shown, particularly in those regions where there is a wide range in elevation, as in the Mountains and Basins. Plants will be found growing naturally where conditions are most suitable for their growth. For example, water lovers will be found in the arid west, but only where there is a dependable supply of water, as along a stream. The paper birch and the claretcup cactus are both adapted to the West Coast region, though neither is native there. However, they will not grow together; the birch needing the cool summers of the north, while the cactus prefers the warm winters of the south. It is always best to check the regional map, the hardiness zones, and the water needs to determine if a particular plant will work in your garden.

PLANT CHOICES

The plants discussed in The Backyard Habitat have been chosen to give an appropriate representation of plants native to each region. In the Northeast and South-east, where forests are the dominant natural climax vegetation type, trees are presented in greater numbers than in the Prairies or Desert Southwest, where native trees are relatively few and climax vegetation is dominated by grasses or shrubs.

PURPLE CONEFLOWER (Echinacea purpurea) *(top); the distinctive giant swallowtail butterfly (above); natural desert garden features mesquite, penstemons, and verbena (right).*

EACH SPECIES will have a natural range within its specific region and will not automatically cover that entire region.

In all regions, a broad selection of plant types has been included to serve the needs of those starting a new garden or adding to existing gardens. Gardeners with any type of garden, be it big or small, damp or dry, sunny or shady, should be able to find plants that suit their circumstances and will attract wildlife to the garden.

KNOW YOUR REGION

The West Coast is a narrow band noted for dryer summers and wetter winters.

The Mountains and Basins include the "cold desert" of the Great Basin states and the adjoining, relatively arid, mountains from the Rockies to the eastern Sierra Nevadas and Cascades.

The Desert Southwest is limited to the warmer Mojave, Sonora, and Chihuahua Deserts to the south.

The Prairies follow the original distribution of the grasslands of the Great Plains.

The Northeast includes the forests and meadows of colder portions of the eastern states.

The Southeast represents the hot, humid, and relatively low-lying lands of the eastern states and the Gulf Coast.

USING *the* BACKYARD HABITAT

Presented on the following pages are 144 native plants that will attract wildlife to the natural garden. They are grouped into six regions: West Coast, Mountains and Basins, Desert Southwest, Prairies, Northeast, and Southeast. A chart at the end of each section lists 20 additional species.

The **common and scientific name** of the plant. Plants are grouped first by type and then in alphabetical order by scientific name within each group.

The **color banding** identifies each of the six regions of The Backyard Habitat.

The **family** in which the plant is placed; and its type (cacti, grass, herbaceous flower, shrub, tree, or vine).

A **photo** of the species discussed; captions are included if the text discusses more than one species.

The **map** shows the plant's native region(s) in green. Any other region(s) to which it has become adapted are shown in orange.

The **text** provides important information on the plant's flowers, foliage, and fruits, including color and season; the preferred soil and propagation method; and which visitors are attracted to the plant, for what reason, and when.

The **calendar bar** highlights the best periods for dividing, taking cuttings, transplanting, or starting seeds.

Accurate, full-color **illustrations** highlight aspects of the flower and fruit, or show visitors to the plant; the sex is noted for birds.

Asteraceae: Sunflowers ✦ Herbaceous flower

Maximilian's Sunflower
Helianthus maximiliani

Maximilian's sunflower makes a real show in the fall landscape. It blooms late, often into October, and the numerous stems lined with many bright yellow flowers create quite a display. It is named after Maximilian Alexander Philipp of Wied-Neuwied, who collected many plants on an expedition up the Missouri River in 1833 and 1834. This is a long-lived perennial that is considered native throughout the Great Plains, but it is most associated with the tallgrass prairies, where there is relatively ample moisture. In dry locations, this sunflower is much smaller than on moist sites. It has become a significant part of northern New Mexico landscaping, where it adds to the dramatic, yellow fall flowers of rabbitbrush (*Chrysothamnus* spp.), various senecios, and the wonderful blue of Bigelow's asters (*Machaeranthera bigelovii*).

Maximilian's sunflower is adapted to a wide range of soils. Propagation is usually by seed. The flowers attract numerous butterflies. Colorful lazuli buntings and white-crowned sparrows feed on the seeds, which have been grown as a source of excellent birdseed.

J F M A M J J A S O N D

FIELD NOTES
- Zones 2–8
- Perennial
- 5–10' (1.5–3 m)
- ☀ Full sun
- ◊ Moderate

white-crowned sparrow

198

Quick-reference Field Notes include:
- The USDA hardiness zone range in which the plant performs best (from coldest to mildest zone).
- Annual or perennial (if the plant is herbaceous).
- Deciduous, semi-evergreen, or evergreen (if the plant is woody).
- The height of the plant.
- ☀ Sun or shade preferences of the plant.

◊ The plant's normal water needs, once established—within its native range.
Low—little or no water beyond natural rainfall.
Medium—occasional water, especially when rainfall is lighter than normal for your area.
High—frequent water, except when rainfall is higher than normal for your area.

The West Coast

THE WEST COAST
The California Dry-summer Garden

The West Coast is comprised of a long region from the crest of the Sierra Nevada and Cascade ranges, in the east, to the Pacific Ocean, in the west, and from the border with Baja California, in the south, to the coastline of Alaska, to the north. Thus defined, the region embraces extremes in climate from short growing seasons with extreme winter cold, to mild, subtropical conditions with a nearly continuous growing season and little frost.

The population is concentrated in the milder coastal zones that experience relatively minor temperature fluctuations throughout the year. The greatest concentration of people is, in fact, in the Mediterranean climate of California, where mild winters and springs represent the rainy season; and dry, warm to hot summers, the drought season.

Xeriscape is the key word, meaning a garden that needs little or no supplemental water. California xeriscape gardens feature a multitude of small trees and flowering shrubs that enliven the scene in spring and attract a wide array of colorful pollinators—honey-, bumble-, and solitary native bees; wasps; hover and bee-flies; butterflies; and hummingbirds. They also offer shelter and housing to tree-dwellers.

The California garden showcases spring-flowering perennials, annuals, bulbs, and native bunchgrasses, which are important to pollinators through a long and varied spring, starting in late February and continuing to early June. This is followed by a quiet time, when many flowers die back to underground roots, and perennial grasses provide texture and color with their leaves. Seeds and seedpods attract various wildlife, such as the many kinds of seed-eating birds and forest mammals, including squirrels and chipmunks. Grasses also provide nesting sites for ground-dwelling birds.

Most shrubs are evergreen but grow little in summer. Some offer colorful, fleshy fruits in summer and fall, which are irresistible to fruit-feeding birds and small mammals. Shrub foliage is also attractive to deer, and it may be necessary to take protective measures, or the growth of the plant will be jeopardized.

A small pond can serve as a precious water supply. Ponds are places of quiet beauty; a periphery for wildflowers that prefer to have wet "feet"; and a welcome source of sustenance to birds, butterflies, and small mammals. They are effective in attracting wildlife to the garden during the day as well as night.

Twining brodiaea (Dichelostemma volubile)

Brodiaeas

Brodiaea, Dichelostemma, and Triteleia *spp*.

Easily adapted to summer-dry gardens, these western natives grow from bulb-like corms, and multiply readily by means of cormlets. Their colorful flowers appear through the spring months; by selecting several kinds you can keep the floral pageant going from March to mid-June. Flowers come in a wide variety of colors: blue, purple, white, yellow, pink, and red with green (firecracker brodiaea, *Dichelostemma ida-maia*), with most being a shade of blue or purple. Seeds spill from seedpods in May to July. Each plant has two to four long, lance-shaped leaves, which often shrivel by flowering time.

Brodiaeas can be planted throughout an open meadow or border. They are best planted in drifts for floral impact. Propagation is from seed (which takes three to four years to bloom) or cormlets. Soil should be a well-drained loam, although sand is best.

The flowers lure a wide variety of beautiful butterflies, especially the spectacular swallowtails, and also bees and bee-flies. Pocket gophers are attracted to the edible corms. It is wise to take protective measures, such as placing corms in wire baskets.

| J | F | M | A | M | J | J | A | S | O | N | D |

anise swallowtail butterfly

FIELD NOTES
■ *Zones 7–10*
(a few to zones 5–6)
■ *Perennial*
■ *2"–2' (5–60 cm)*
☀ *Most like full sun; some accept light shade*
🌢 *Low*

108

Douglas Iris

Iris douglasiana

The widely spreading clumps of glossy, deep green, sword-shaped leaves look good all year. Douglas iris's rhizomatous colonies should be divided every three years or so. The delicate, orchid-like blossoms light up meadows in early to mid-spring. Flower color varies from lavender to deep blue, white to pale yellow, and flowering occurs at different times, according to locale. By selecting several forms you may extend garden bloom. Many hybrids are available, but many of these are not so drought tolerant nor as delicate as the true species. Elongated fruits ripen about three weeks after the flowers bloom.

J F M A M J J A S O N D

Propagation is by root division and seed (stratification may improve the percentage of germination). Douglas iris prefers a well-drained soil, and it is tolerant of rocky soil. Plant it throughout a meadow or toward the front of a mixed border. It will tolerate light shade. Bees, bumblebees, bee-flies, checkerspot, copper, and swallowtail butterflies, and Costa's and Anna's hummingbirds feed on the flower nectar. The rhizomes and leaves are poisonous to livestock and not attractive to deer.

seedpods

FIELD NOTES
■ Zones 8–10
■ Perennial (evergreen)
■ 8–18" (20–45 cm)
☀ Full sun, especially near coast; light shade elsewhere
⬥ Low

bumblebee

109

Leopard Lily

Lilium pardalinum

The graceful stems of leopard lily emerge in spring, and slowly grow to several feet tall, carrying narrow, smooth-edged, spirally arranged leaves. The flower buds swell and open toward spring's end, revealing large, nodding (or hanging), orange or red-orange with yellow-centered blossoms, with beautifully recurved petals sprinkled with dark brown or black-purple spots, such as those on a leopard's coat. The fruits ripen about a month after flowering. Papery seedpods appear in mid- to late summer.

Propagation is by bulb scales, divisions, or seed, but the plants will take three or more years to reach blooming size.

J F M A M J J A S O N D

Leopard lily prefers a well-drained soil, with the ability to retain moisture. A perfect position for leopard lily is along the periphery of a stream or pond, in light shade. Plant the bulbs between rocks.

papery seedpods

Leopard lily flowers attract beautiful pollinators, such as large butterflies (including monarchs, admirals, and mourning cloaks), sphinx moths, Allen's and rufous hummingbirds, and bumblebees. Bulbs are eagerly sought by pocket gophers. Deer love to browse the stems and especially the flower buds, so it may be necessary to protect your plants with wire cages.

FIELD NOTES
- Zones 5–10
- Perennial
- 3–6' (1–1.8 m)
- ☀ Full sun near coast—filtered sun inland
- ◌ Moderate

mourning cloak butterfly

Lupinus arboreus

Lupines

Lupinus spp.

Lupines range from annual to perennial to small shrub, according to kind. All are characterized by their palmately compound leaves, the leaflets splayed out like the fingers on a hand. The leaves are sometimes covered with silky to felted white hairs. The showy, pea-like flowers are arranged in whorls or spirals up long spikes and come in many beautiful colors, including cream and white, with blues and purples predominating. The harlequin lupine (*L. stiversii*) is striking for its pink and bright yellow color combination. Lupine roots bear nodules that add nitrogen to the soil after the plants have died, helping to enrich impoverished soils. Lupines all lend beautiful color and form to the landscape from early spring well into summer. Choose several species to prolong your floral pageant.

Propagation is from seed, which should be soaked and/or stratified, particularly if the seed is not fresh. Lupines prefer a sandy or well-drained soil. Meadow lupine (*L. polyphyllus*) requires a moist soil. Nonwoody lupines are best massed in a meadow or border; shrubby lupines can be planted in front of taller shrubs.

| J | F | M | A | M | J | J | A | S | O | N | D |

Bumblebees and a wide variety of butterflies, such as the common hairstreak and acmon blue, are attracted to the often-fragrant lupine flowers. Over 40 different species of lupine provide larval food for the common blue butterfly.

common blue butterfly

FIELD NOTES

■ Zones 3–10 for perennials; 5–10 for shrubby species; all zones for annuals

■ Annual or perennial

■ Evergreen

■ 4"–5' (10–150 cm)

☀ Most need full sun

◐ Low–moderate

Scarlet Monkeyflower

Mimulus cardinalis

Scarlet monkeyflower grows along stream banks. Easy to grow, its rather floppy, flowering branches bear a long succession of scarlet-red blossoms in summer and fall. The flowers look as though they are yawning, for the two lips of the petals are spread in such a way that the lower lip is curved backward and the upper lip stands straight up. This arrangement, together with the vivid flame-colored blossoms, make it an excellent hummingbird-pollinated flower. Opposite, pale green leaves grow up the stems.

Propagation is by root division or from seed. Scarlet monkeyflower prefers sandy or well-drained soil and tolerates rocky areas.

J F M A M J J A S O N D

Plant beautiful clumps by a water feature, in light shade. Old flower stalks should be cut back to promote new flowering. Seedpods follow three weeks after flowering.

Scarlet monkeyflower blooms are very attractive to Anna's, Costa's, Allen's, and rufous hummingbirds, especially late in the summer season. Common checkerspot and buckeye butterflies lay their eggs on the leaves.

rufous hummingbird ♂

FIELD NOTES
- ■ Zones 6–10
- ■ Short-lived perennial (2–3 years)
- ■ 2–4' (60–120 cm)
- ☀ Light shade is best
- ◊ High

112

Foothill Penstemon

Penstemon heterophyllus

Of the numerous ornamental penstemons available, foot-hill penstemon is one of the best. It grows rapidly, blooms prolifically from late spring through summer, and adapts readily to gardens. The several leafy stems carry long wands of yellow buds and showy blue to purple flowers. Nearly linear, smooth-edged, pale green to whitish-green leaves grow up the stems. Flower color may change with age and from plant to plant; sometimes, there are tints of rose-purple, at other times, clear blue. Foothill penstemon blooms longer when it receives summer water, but with a shortened lifespan. This does not matter, for the plants are easily propagated from tip cuttings (with some older, woodier growth at the base desirable), or from seed. Seedpods ripen three weeks or so after the flowers finish blooming. Foothill penstemon is excellent in rock gardens, at the front of mixed borders, or between bunchgrasses in meadows. It likes a well-drained soil, and sandy or rocky soil is fine.

A wide range of pollinators visits the flowers, including native bees, bee-flies, wasps, and hummingbirds.

native bee

| J | F | M | A | M | J | J | A | S | O | N | D |

digger wasp

FIELD NOTES

■ Zones 6–10; protective mulch needed in areas with coldest winters

■ Short-lived perennial (2–3 years)

■ Evergreen

■ 1–2½' (30–75 cm)

☀ Full sun

◊ Low

113

California Fuchsia

Zauschneria spp.

California or hummingbird fuchsia is a subshrubby perennial with widely traveling roots. Its stems bear myriad, narrow, pale green to grayish-white leaves; plants may be upright and bushy or prostrate and groundcovering. The trumpet-shaped, scarlet flowers appear from midsummer through fall, depending when the first hard frosts occur. Pink and white cultivars are also available. The flowers are followed two weeks later by fruits, which are long, slender capsules that split into four segments to release numerous seeds covered with tufts of white hair. California fuchsia needs to be cut back to the ground when it goes dormant in late fall, which makes it a great companion to early spring flowers that will be finished by the time *Zauschneria* gets going.

Propagation is from seed (which may self-sow), or by root divisions. Sandy or rocky soils are ideal for California fuchsia. Plant it where there is plenty of room for it to spread, such as through a wildflower meadow. California fuchsia's chief charm is the long succession of scarlet flowers that are irresistible to all kinds of hummingbird, providing them with a regular source of food in late summer and fall.

J F M A M J J A S O N D

Anna's hummingbird

FIELD NOTES
- Zones 6–10
- Perennial
- Winter dormant
- 6–18" (15–45 cm) tall
- ☀ Full sun
- ◐ Low

Sonoma manzanita (*Arctostaphylos densiflora* "Sentinel")

Manzanita

Arctostaphylos spp.

Manzanitas fill several niches in the garden because of their variety of forms. Kinnikinnick (*A. uva-ursi*) provides a drought-tolerant evergreen groundcover, while the bigberry manzanita (*A. glauca*) may grow to a small tree. Plant manzanitas in the background of the garden, with other drought-tolerant shrubs, but not in the shade of trees. Groundcover species are fine in light shade from shrubs or small trees. All manzanitas feature smooth to flaky, polished, red-purple bark and tough, evergreen leaves (which may vary from silvery white to vivid green or gray-green) Masses of nodding, urn-shaped, white or pink flowers bloom from winter through early spring, according to kind. The late spring through summer fruits resemble little apples, hence the common name (manzanita is Spanish for "little apples").

Propagation is from scarified or stratified seed. Cuttings may be taken in fall using bottom heat. Manzanita favors sandy to loamy soil, but with excellent drainage. Some species prefer acid soil.

A wide range of animals, from bears and coyotes to fruit-eating birds (such as California thrashers, rufous-sided towhees, and fox sparrows) are fond of the fruits; butterflies, bee-flies, flies, and small bees frequent the flowers from late winter to early spring.

J	F	M	A	M	J	J	A	S	O	N	D

winter–spring flowers

fox sparrow ♂

FIELD NOTES

■ Zones 5–10, according to species

■ Evergreen

■ 4"–20' (10 cm–6 m)

☀ Most want full sun (strictly coastal species prefer light shade)

◊ Low

Santa Barbara ceanothus (*Ceanothus impressus*)

Wild Lilacs

Ceanothus *spp.*

Like the manzanitas, wild lilacs come in a wide range of forms, from creeping, woody groundcovers to small multitrunked trees. Most have drought-resistant, evergreen leaves. Flowers bloom from mid-March to late May or early June, according to species, and are arranged in dense clusters of blue, purple, pinkish, or snowy white. The smell of the flowers on warm days is irresistible, reminiscent of honey and corn tortillas. The three-sided seedpods ripen in May to July. They start off shiny and fleshy and are often decorative, but eventually dry out and split open.

fruits of C. cordulatus

Propagation is from semi- to hardwood cuttings in fall or stratified seed. Plant in well-drained soil; rocky soil is often acceptable. Most wild lilacs are best planted with other drought-tolerant shrubs; a few tolerate the shade of adjacent trees, including deerbrush (*C. integerrimus*) and white-thorn (*C. incanus*).

The blossoms are like magnets to bees and butterflies. Deer find the young foliage attractive. Large, shrubby kinds provide habitat for ground-nesting birds. Many kinds provide food for the larvae of the pale swallowtail butterfly.

| J | F | M | A | M | J | J | A | S | O | N | D |

pale swallowtail butterfly

FIELD NOTES
- Zones 5–10, with most in zones 7–10
- Evergreen (except for C. integerrimus)
- 2"–20' (5 cm–6 m),
- ☀ Most prefer full sun, a few light shade
- ◐ Low

116

Sulfur buckwheat (*Eriogonum umbellatum*)

Wild Buckwheats

Eriogonum spp.

The wild buckwheats comprise a decorative and populous clan of annual, perennial, or shrubby plants beautifully adapted to summer-dry conditions. Long-lasting, spoon-shaped to narrow, evergreen leaves are covered beneath with a thick wool. Buckwheats bloom from late spring into fall, according to kind, and provide light colors—whites, pinks, cream, and yellows—often fading to a colorful red or rusty brown. Sulfur buckwheat (*E. umbellatum*) has bright yellow flowers that fade to red. One-seeded fruits follow two to three weeks after the flowers.

Propagation is from semihardwood cuttings or seed. Buckwheats thrive in sandy, loamy, or rocky soil with good drainage. Plant them in full sun in the border, in a rock garden, or in an open, rocky meadow.

| J | F | M | A | M | J | J | A | S | O | N | D |

Long-horn beetles, bee-flies, wasps, butterflies, and various bees are attracted to the blossoms; gold-finches and other seed-eating birds seek out the fruits; and bramble green hairstreak, blue copper, and metalmark butterflies lay their eggs on leaves for their larvae to feed on.

bramble green hairstreak caterpillar

FIELD NOTES
- Zones 2–10
- Evergreen for perennial and shrubby species
- 3"–4' (8–120 cm)
- ☀ Full sun for most species
- 💧 Low

Salal

Gaultheria shallon

The semiprostrate to upright, zigzaggy stems of salal weave intricate shrubberies and hedges along the edge of coastal forests. They will do likewise in the garden, enlarging their territory by means of their wandering roots. The handsome, large, evergreen leaves are stiff and decorative. Modest clusters of hanging, bell-shaped, white to pale pink blossoms, with pink sepals, appear in May to July, followed in July to September by large, near-black, edible "berries".

J F M A M J J A S O N D

Propagation is by root division and seed, which should be stratified for one to two months. Well-drained acidic loams and sands, conditions as for rhododendrons, camellias, or azaleas, best suit salal. Plant salal as an informal hedge or wandering shrub along a forest border; it is best planted under conifers, such as Douglas fir (*Pseudotsuga menziesii*).

summer fruits

A wide variety of fruit-eating birds, bears, and small rodents, such as Sonoma and Townsend chipmunks, eat the fruits. Various small bees, flies, and hover flies are attracted to the flower nectar. Browsing mule deer can be a pest and you may need to take steps to protect the plant.

mule deer

FIELD NOTES

- ■ *Zones 7–10*
- ■ *Evergreen*
- ■ *1–10' (30–300 cm)*
- ☀ *Moderate–full shade*
- ◐ *Moderate–high*

Toyon

Heteromeles arbutifolia

Toyon, or California holly, grows into a large shrub or small tree in well-drained soils and with hot summer temperatures. The tough, serrated, dark green leaves are attractive and a good replacement for true holly (*Ilex aquifolium*), as are the brightly hued, red-orange berries that ripen in late fall. Berries sometimes remain on the plant through winter and early spring, when they feed northward-bound migrating birds. The rounded to pyramidal clusters of tiny, whitish flowers appear at the end of spring.

Toyon is best planted at the back of a meadow or border, with other drought-tolerant shrubs, or in front of oaks. It prefers a well-drained soil. Propagation is from semihardwood cuttings (using bottom heat). Seed should be stratified for one to two months.

| J | F | M | A | M | J | J | A | S | O | N | D |

Fall to winter resident birds, such as yellow-rumped warblers, California thrashers, starlings, rufous-sided towhees, and house sparrows are attracted to toyon in fall. Beetles, wasps, hover flies, butterflies, and bees are attracted in early summer to the abundance of nectar in the masses of tiny blossoms.

California thrasher ♂

FIELD NOTES

◾ Zones 6–10

◾ Evergreen

◾ 10–30' (3–9 m)

☀ Full sun–light shade

◐ Low

California Coffeeberry

Rhamnus californica

California coffeeberry is an easy to grow, densely branched shrub ranging from coastal forms with sprawling branches, which may be wind-pruned to less than a foot tall; to inland shrubs with many upright branches, which may grow to over 10' (3 m). The attractive leaves vary from deep green to a dusky, pale green according to location. Tiny, greenish-yellow, star-like blossoms are hardly noticed in late spring, but the fleshy berries add interest as they turn from green to red to deep purple in early fall.

Propagation is by layering, from hardwood cuttings and seed (which should be stratified for one to two months). California coffeeberry prefers a sandy or well-drained soil, and requires minimal water. It can be placed in front of taller shrubs in the garden, in company with smaller manzanitas and ceanothuses.

J F M A M J J A S O N D

Fruit-eating birds, such as cedar waxwings, robins, and black-headed grosbeaks, seek out the berries, and wasps, beetles, and hover flies visit the flowers to feed on their nectar. The dense branches provide cover for cottontail and other rabbits, and deer are fond of the young leaves. The pale swallowtail butterfly lays its eggs on the leaves.

fall fruits

FIELD NOTES

■ Zones 5–10

■ Evergreen

■ 1–12' (30 cm–4 m)

☀ Full sun–light shade

💧 Low

pale swallowtail caterpillar

Pink-flowering Currant

Ribes sanguineum *var.* glutinosum

Pink-flowering currant is a medium-sized shrub that grows rapidly to maturity in partial shade on the border of woods. Its small, pale green, maple-like leaves are sage-scented on warm days, but its best feature is the numerous trusses of hanging deep rose, pale pink, or white blossoms at winter's end. Purplish berries follow in summer.

J F M A M J J A S O N D

hanging blossoms

Propagation is from cuttings in winter and early spring. It prefers a well-drained soil, but will tolerate slightly acid or neutral soil. Pink-flowering currant is ideal planted as a backdrop in a garden with light shade.

Fruit-eating birds, including yellow-rumped warblers, California thrashers, house sparrows, and starlings, and small mammals, eat the fruit in late summer and early fall, although the berries are unlikely to be their first choice when other berries are available. Colorful pollinators such as hummingbirds and bees visit the flowers for their nectar in early spring. The silenus anglewing butterfly lays its eggs on the leaves. Mule deer browse the branches, especially the new growth in early spring.

summer fruits

FIELD NOTES
- Zones 7–10
- Deciduous
- 6–12' (2–4 m)
- Best in light shade
- Low–moderate

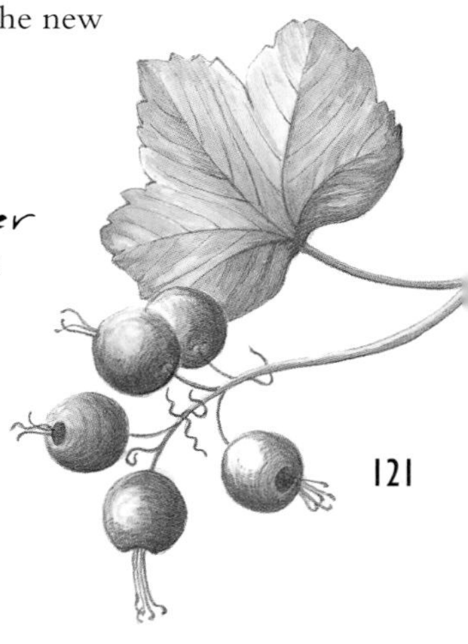

121

Thimbleberry

Rubus parviflorus

Thimbleberry is a widely ranging, small to medium-sized shrub with wandering roots. Like its close sisters the black-berries and raspberries (*Rubus* spp.), it may become invasive, but forms natural, good-looking groupings and loose hedges in a suitable, spacious woodland setting. Large, pale green, downy-soft, maple-like leaves complement the single, rose-like white blossoms in May to July. Thimble-shaped, dark red berries follow in late June to September, with an exquisite taste reminiscent of raspberries.

Propagation is by root division or from seed (which must be stratified). Thimbleberry

spring flower

| J | F | M | A | M | J | J | A | S | O | N | D |

prefers a well-drained, acid soil. Plant it along the edge of woods. It prefers a shady, garden position.

Bees, beetles, and flies visit flowers in late spring and early summer. Squirrels and fruit-eating birds, such as Swainson's and hermit thrushes and black-capped vireos, avidly seek the fruit in summer or early fall. Mule deer eat the tender new growth in early spring, so make sure you take measures to protect the plants.

Swainson's thrush ♂

FIELD NOTES
- Zones 5–10
- Deciduous
- 4–10' (1.2–3 m)
- Light–fairly deep shade
- Low–moderate

White sage (*Salvia apiana*)

Sages

Salvia spp.

The sages fill a key role in dry gardens, through their drought tolerance, late spring blooming, scented leaves, and colorful flowers. Sages range from annual thistle sage (*S. carduacea*), with elegant, ruffled, pale blue flowers, to the shrubby sages, such as white sage (*S. apiana*), with broadly elliptical, whitish leaves and tall spires of white to palest purple blossoms. All have strongly scented leaves; tiers of whorled, two-lipped blossoms; and decorative seedheads that dry on the plant. Leaf color varies from whitish to pale green or gray-green. A number of Central American species of *Salvia* have become popular garden plants across North America.

Propagation is from half-woody cuttings or seed (which should be soaked first). Sages prefer a well-drained soil; sandy to rocky is best. Plant them in the garden in full sun, as a component of a dry meadow (annual kinds), or in the middle ground next to taller shrubs (woody kinds).

Hummingbirds, bees, bumblebees, and bee-flies seek the flower nectar in spring and summer. Goldfinches, house sparrows, and quail eat the fruits (seeds). Many sage species are considered partially deer-proof, as deer don't like the highly scented leaves.

| J | F | M | A | M | J | J | A | S | O | N | D |

FIELD NOTES
- Zones 4–10, according to kind (sow seeds of annual kinds in any zone after danger of frost has passed)
- Evergreen
- 3–6' (1–1.8 m)
- ☀ Most prefer full sun; hummingbird sage (*S. spathacea*) takes light shade
- 💧 Low

honeybee

123

Blue Elderberry

Sambucus mexicana

lue elderberry is a fast-growing, graceful, large shrub or small tree. If suckers are pruned out, either one or a few main trunks may develop. The grooved bark; large leaves, which give off a strong odor when crushed; and flat-topped clusters of numerous tiny, yellowish-white blossoms distinguish blue elderberry from any other plant. The spring blossoms are fragrant up close, emitting a stronger perfume in the evening to lure night-flying insects for pollination. Strongly flavored, bluish berries follow in early summer.

J F M A M J J A S O N D

Propagation is by means of rooted suckers, semihardwood cuttings, or seed (which should be stratified for a couple of months). Elderberry prefers a well-drained soil, but a high water table is best so roots can go deep to find water under summer conditions. Plant blue elderberry in the back of the garden in a low spot if the soil drains well. Also plant it next to a water feature but not directly in soggy soils.

Long-horn beetles, sawflies, butterflies, and bees all visit the flowers in late spring to feed on their copious nectar. Many fruit-eating birds relish the ripe summer fruits, particularly band-tailed pigeons and house finches.

FIELD NOTES
■ Zones 2–10
(coldest climate may require some protection)
■ Deciduous
■ 10–45' (3–14 m)
☀ Full sun–light shade
◐ Moderate

viceroy butterfly

124

Evergreen Huckleberry

Vaccinium ovatum

Evergreen huckleberry is a relatively slow-growing, medium-sized, compact, and handsome evergreen shrub that favors the understory of coniferous forests along the coast. The small, glossy leaves are useful for Christmas greens; the branches are laden with pale pink to white bells in mid- to late spring; and the branches carry luscious, dark purple to pale bluish berries in summer.

Plant evergreen huckleberry so that you can look up into the branches from below, as the flowers hide underneath the branches. Evergreen huckleberry is best planted along the edge of a forest or woodland as it needs shade. In the garden, plant it in the partial shade of a well-established, preferably evergreen, tree.

Propagation is from hardwood cuttings (using bottom heat), or seed (which should be stratified for a couple of months). It prefers a well-drained and acid soil.

Butterflies and several kinds of bee visit and pollinate the flowers. The delicious fruits are eaten by many birds, including western tanagers, white-crowned sparrows, and orange-crowned warblers, as well as squirrels, chipmunks, and bears.

J F M A M J J A S O N D

western tanager ♂

FIELD NOTES
- Zones 7–10
- Evergreen
- 4–10' (1.2–3 m)
- Light–medium shade
- Moderate

125

California Buckeye

Aesculus californica

California buckeye provides a handsome backdrop to the garden, by itself or with other small trees. It is a beautifully rounded tree with seasonal interest all year. Leafless from late summer through winter, its smooth, silvery white bark is highlighted then. The new flush of apple-green leaves is a welcome event in early spring, and candles of white to pale pinkish flowers illuminate branch tips in late May or early June. Leathery seedpods split open in fall to reveal one large, shiny brown, chestnut-like seed.

To propagate, plant fresh seed in well-drained soil in late fall to early winter. Buckeye's name derives from the resemblance of the open seedpod to a large eye. Despite their resemblance to chestnuts, buckeye seeds are deadly poisonous to humans.

Buckeye flowers attract all manner of beautiful butterflies (such as the Chalcedon checkerspot), sphinx moths, and hummingbirds. The nectar is poisonous to bees. Branches provide perches and nesting sites for birds, including yellow-billed cuckoos and downy woodpeckers, although the nests are readily apparent in winter. The large seeds attract small mammals which are not affected by the toxicity.

J F M A M J J A S O N D

seedpods

Chalcedon checkerspot butterfly

FIELD NOTES
- Zones 7–10
- Deciduous
- 10–30' (3–9 m)
- Full sun–very light shade
- Low–moderate

126

Madrone

Arbutus menziesii

Madrone is a sturdy evergreen tree whose crown seeks light by searching for breaks in the forest cover. A large tree, it is best planted in the background of the garden, on the level, or on a raised mound. The new bark is a striking, glossy red-orange, while the older bark crumbles into shingle-like scales. The tough, leathery leaves are silvery on their backs. White, fragrant blossoms, similar in appearance to those of lily-of-the-valley (*Convallaria majus*), appear in mid-spring. Round, red-orange, wart-covered berries appear in late fall.

This is a difficult tree to transplant, and seed should be stratified for a couple of months before planting. Plant madrone in a well-drained loam; a mildly acid soil is preferable. It may require extra water in hot, exposed situations.

The branches of madrone provide an excellent habitat for tree-nesting birds, while the crown is a good lookout post for hunting birds. Fruit-eating birds, such as cedar waxwings and varied thrushes, and also coyotes and squirrels, flock to the berries. Bees and bee-flies are fond of the flower nectar.

J F M A M J J A S O N D

varied thrush ♂

FIELD NOTES

■ Zones 7–10

■ Evergreen

■ 100' (30 m)

☀ Full sun–light shade

◊ Low

Holly-leaf Cherry

Prunus ilicifolia

Holly-leaf cherry, or islay, grows as a dense shrub, which is prunable into a hedge, or into a small, multitrunked tree. The glossy, bright green foliage is holly-like, down to the sharply toothed leaf margins. Modest racemes of whitish flowers decorate branch tips in late spring, followed by large, plump, red-purple cherries in late summer. This shrub is ideal as a backdrop to a flower border or meadow. Plant it at the back of the garden, perhaps with other large shrubs, such as toyon (*Heteromeles arbutifolia*) and manzanitas.

J F M A M J J A S O N D

Propagation is from seed (stratify briefly) or hardwood cuttings using bottom heat. Holly-leaf cherry prefers a well-drained soil; a rocky or sandy soil is suitable.

Many bees, beetles, and hover flies feed on the blossoms' nectar. Two-tailed swallowtail and Lorquin's admiral butterflies sometimes lay their eggs on the leaves. Various fruit-eating birds, such as house finches, and small mammals, such as gray squirrels and chipmunks, relish the fruits.

FIELD NOTES

- Zones 7–10
- Evergreen
- 50' (15 m)
- Full sun
- Low

♂

house finch

128

Douglas Fir

Pseudotsuga menziesii

Douglas fir is one of the West's truly magnificent trees. Crowns may exceed 200' (60 m) and trees may live for several hundred years. The mature bark is craggy and irregularly split; branches carry deep green, spirally arranged needles with a wonderful pine fragrance; new needles are soft and bright green. Decorative seed cones drip from branch tips, ripening from green to brown in late fall.

Propagation is from seed. Douglas fir prefers a well-drained, acid soil.

seed cone

| J | F | M | A | M | J | J | A | S | O | N | D |

Perfect for the backdrop of a large garden, Douglas fir can be planted with compatible trees, such as madrone (*Arbutus menziesii*), tanbark oak (*Lithocarpus densiflorus*), bay-laurel (*Umbellularia californica*), and coast redwood (*Sequoia sempervirens*).

Because of Douglas fir's size, many animals shelter under its lower branches, while tree-nesting birds favor higher branches, and osprey, eagles, and hawks perch on the tree's top to watch for prey. Bald eagles, band-tailed pigeons, hummingbirds, western-wood peewees, and Steller's jays nest in Douglas fir, and ospreys nest on dead snags. Raccoons are known to nest in trunk cavities; Douglas' and gray squirrels eat the seeds. The pine white butterfly's larvae eat the needles.

FIELD NOTES

■ Zones 2–10; select locally grown plants for greatest adaptability

■ Evergreen

■ 50–200'+ (15–60 m)

☀ Full sun–light shade

◐ Low–moderate

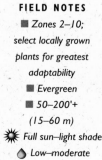

Douglas' squirrel

129

Coast Live Oak

Quercus agrifolia

The rounded canopy, large muscular trunk and primary limbs, and dense shade, typify mature coast live oaks. The leaves are glossy and evergreen and distinctive because their margins curl under, especially where leaves are subjected to bright sun.

J F M A M J J A S O N D

Sharp prickles line leaf edges. The long, dangling male catkins, reddish then becoming yellow on opening, decorate branches in early spring, while the female blossoms are tiny and seldom noticed. The fall acorns are easy to see and vary in abundance from year to year. Each acorn sits in a scaly cup and is long and pointed, often decorated with darker brown stripes.

Propagation is from seed (acorns), which should be sown when fresh. Coast live oak likes a well-drained soil. Plant this tree at the back of a garden, and allow ample room for the full development of the crown.

Great horned owls, red-tailed hawks, scrub jays, bushtits, and American robins, nest in the tree canopies. California sister, tailed copper, and California hairstreak butterflies lay their eggs on the leaves. Gall insects (midges and stingless wasps) cause the formation of diversely shaped, decorative galls on leaves, stems, and branches. Acorn and other woodpeckers, scrub jays, magpies, and several kinds of squirrel and chipmunk eagerly seek the acorns for food.

acorn woodpecker ♂

FIELD NOTES
- Zones 7–10
- Evergreen
- 25–50' (7.5–15 m)
- Full sun–light shade
- Low

130

California Pipevine

Aristolochia californica

California pipevine, also known as Dutchman's pipe, climbs by winding its stems around adjacent shrubs or small trees. If you train it up a tall shrub or young tree, or over a trellis, in partial shade, it will grow rapidly, quickly establishing a covering over branches and festooning them with felted, elongated, heart-shaped leaves. Vines stand out in late winter to early spring (February to early April), when the leafless stems carry the bizarre pipe-shaped, brownish, maroon, and greenish flowers with a maroon mouth. The fruits are 3" (7.5 cm) long, six-angled oval pods that ripen in early summer. The fluted seedpods are decorative in their turn by late spring.

winter-spring flower

J F M A M J J A S O N D

Propagation is from stem cuttings in fall and winter. California pipevine likes well-drained soil, but it will accept a rocky soil. This vine is admired for its unusual flowers.

California pipevine is the exclusive food plant for the iridescent black–purple pipevine swallowtail butterfly. The adults are drawn to the vine at the time of egg laying. The flowers are pollinated by minute mycetophilid gnats, which are temporarily trapped inside the pipes.

FIELD NOTES
- ◼ Zones 7–10
- ◼ Winter deciduous
- ◼ 20' (6 m)
- ☀ Light–medium shade
- 💧 Low–moderate

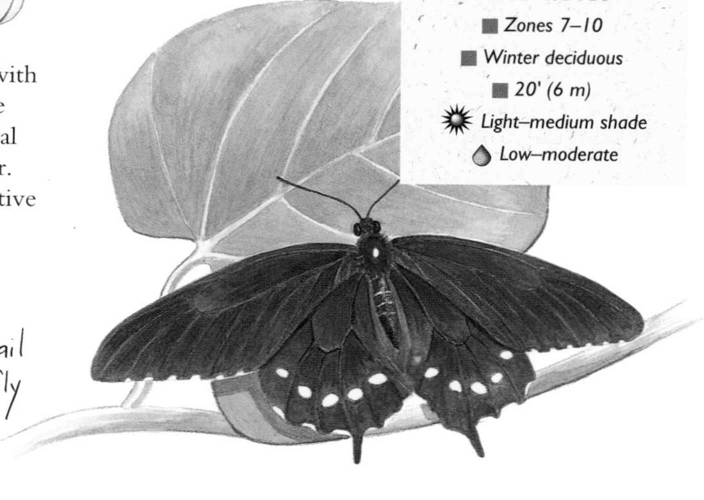

pipevine swallowtail butterfly

131

Rhododendron (left); Cape honeysuckle (right)

Additional species

Plant name/flower color	Zone	Height	Visitors/attraction/season
GRASS Purple needlegrass (green, gold) *Stipa pulchra*	5–10	3–4' (90–120 cm)	birds/seeds/summer–fall
Wild onions (white, purple, pink) *Allium* spp.	wide range	2–12" (5–30 cm)	butterflies/flowers/spring pocket gophers/bulbs/spring
Red columbine (red–orange, yellow) *Aquilegia formosa*	7–10	2–4' (60–120 cm)	hummingbirds/flowers/ late spring and summer
Wallflowers (yellow, orange, cream) *Erysimum* spp.	5–10	1–4' (30–120 cm)	bees, butterflies/flowers/ mid-spring to early summer
California poppy (golden orange) *Eschscholzia californica*	8–10	12–18" (30–45 cm)	bees/flowers/spring
Gilia (blue, lavender, blue–violet) *Gilia* spp.	most	4–24" (10–60 cm)	bees, bumblebees, small butterflies/ flowers/mid- to late spring
Lavender (lavender, purple) *Lavandula* spp. (exotic)	7–10	2–4' (60–120 cm)	bees/flowers/late spring–summer
Hooker's evening primrose *Oenothera hookeri* (clear yellow)	most	2–8' (60–240 cm)	hummingbirds, butterflies, sphinx moths/flowers/summer
California coneflower (golden and black) *Rudbeckia californica*	7–10	3–8' (90–240 cm)	butterflies/flowers/mid- to late summer
SHRUBS Hardy fuchsia (red, purple) *Fuchsia magellanica* (exotic)	6–10	4–6' (1.2–1.8 m)	hummingbirds/flowers/all year
Island snapdragon (clear red) *Galvezia speciosa*	9–10	up to 3' (90 cm)	hummingbirds/flowers/winter–spring
Azaleas and rhododendrons (various) *Rhododendron* spp. and hybrids (exotic)	6–10	5–10' (1.5–3 m)	butterflies, hummingbirds/ flowers/late spring–summer
Fuchsia-flowered gooseberry (bright red) *Ribes speciosum*	9–10	3–6' (90–180 cm)	hummingbirds/flowers/mid-spring fruit-eating birds/berries/summer
Nootka rose (pink) *Rosa nutkana*	6–9	6' (1.8 m)	bees/flowers/spring–summer birds/fruits/fall
Snowberry (pink, white) *Symphoricarpos rivularis*	2–9	6–8' (1.8–2.4 m)	birds/fruits/winter
TREES Vine maple (red, white) *Acer circinatum*	6–9	15–25' (4.5–7.5 m)	birds/nesting/spring birds/seeds/summer–fall
Western hemlock (green) *Tsuga heterophylla*	6–9	100'+ (30 m)+	birds/seeds/fall birds/nesting/spring
California bay laurel (yellow) *Umbellularia californica*	8–10	20–50' (6–15 m)	bees/flowers/winter squirrels/fruits/fall
VINES Cape honeysuckle (orange, yellow) *Tecomaria capensis* (exotic)	9–10	8–12' (2.4–3.7 m)	hummingbirds/flowers/fall–winter
California wild grape (green) *Vitis californica*	8–10	20–30' (6–9 m)	bees/flowers/late spring birds/fruits/mid- to late fall

Mountains and Basins

MOUNTAINS AND BASINS
The Mountain Garden

The region of the Mountains and Basins encompasses a large area of western North America between the West Coast and the Prairies, and extends from northern Arizona and New Mexico north to parts of Alaska. The major geological features are the mountain ranges—most notably, the Rockies. In between are relatively dry valleys and broad basins, such as the Great Basin of Nevada and adjoining states, which is often called a cold desert because of its low annual rainfall and cold winter temperatures. Snow is common in most areas in winter, and showers in summer, but the region is decidedly arid. Lower elevations experience intense heat in the summer, while the high elevations remain cool.

The vegetation varies from grasslands to shrub-dominated basins, to mountain slopes clothed in an open forest of needle-leaf conifers and deciduous aspens. Spring sees a grand display of wildflowers in the lowlands, while summer is the peak season for wildflowers at higher elevations. The growing season shortens as elevation increases.

Where the summers are relatively cool at the middle elevations, the mountain garden might be blessed with an existing open canopy of ponderosa pine (*Pinus ponderosa*), favored by three different species of nuthatch. On a cool north slope is a lively grove of quaking aspen (*Populus tremuloides*). In the slightly more moist soil near its base are drifts of blue Rocky Mountain columbine (*Aquilegia caerulea*), that are popular with hummingbirds. Creeping mahonia (*Mahonia repens*) carpets the ground in shadier areas beneath the trees. On the edge of a seepage, the white-flowered Saskatoon serviceberry (*Amelanchier alnifolia*) ripens its red fruits in June, providing ample food for chipmunks and birds alike. This moist area is alive in midsummer with the tall, loose spikes of fireweed (*Epilobium angustifolium*), abuzz with bees seeking the nectar.

On a dry, sunny slope, the beardlip penstemon (*Penstemon barbatus*) and the cardinal penstemon (*P. cardinalis*) attract the broad-tailed hummingbird, as do great drifts of scarlet gilia (*Ipomopsis aggregata*). Near the house is a small border: blanket flower (*Gaillardia aristata*) and wild white yarrow (*Achillea millefolium* var. *lanulosa*) appeal to the butterflies; wild four o'clock (*Mirabilis multiflora*) draws in the sphinx moths in the early evening; and Indian pink (*Silene laciniata*) attracts the hummingbirds.

Sideoats Grama

Bouteloua curtipendula

Sideoats grama is an attractive, long-lived bunchgrass. The seeds develop on the side of a zigzag stem and give this grass a distinctive appearance. The fall color is an attractive light brown, which is especially eye-catching when backlit by the sun. Sideoats grama is one of several grasses that make up the midgrass prairies. Color in the midgrass prairies is provided by a variety of wild flowers, such as purple prairieclover (*Petalostemon purpureum*), blanket flower (*Gaillardia aristata*), Lewis's flax (*Linum perenne* var. *lewisii*), and Mexican hat coneflower (*Ratibida columnifera*).

Sideoats grama grows in a wide range of soils. Propagate it from seed. Dry conditions favor bunchgrasses like sideoats grama, and excessive moisture may limit good growth. Wet conditions tend to favor sod-forming grasses.

Coyote, foxes, and meadowlarks are a few of the species that make their homes in midgrass meadows. Sideoats grama is also a host plant for the Dakota skipper butterfly.

J F M A M J J A S O N D

western meadowlark ♂

FIELD NOTES
- Zones 3–10
- Perennial
- 1–3' (30–90 cm)
- Full sun
- Low–moderate

Wild White Yarrow

Achillea millefolium *var.* lanulosa

Wild white yarrow is a meadow wildflower with tiny flowers in flat clusters and finely divided, fragrant leaves. The flowers bloom from May to September, depending on the altitude and climate. This clump-forming plant is quite common in moderately moist areas in the ponderosa pine (*Pinus ponderosa*) forests. Nosebleed, sneezewort, knight's milfoil, bloodwort, and soldier's woundwort are a few of the names that refer to the healing nature of this plant. Some 58 different Native American groups may have used this yarrow for various medicinal treatments, and treatment of coughs, throat irritation, and bleeding were among the most common.

Wild white yarrow grows well in most soils; in fact, when there is ample moisture, this plant can sometimes spread somewhat too vigorously. Propagate it from seed or by dividing established plants.

spring-summer flowers

J F M A M J J A S O N D

Wild white yarrow is a great nectar plant, attracting a variety of butterflies, including the American copper. Deer rarely browse this plant—a very useful quality for an increasing number of gardeners—which may be because of the fragrant foliage or the taste.

FIELD NOTES
- Zones 3–9
- Perennial
- 1–2' (30–60 cm)
- ☀ Full sun or part shade
- ◊ Moderate

American copper butterfly

137

Mosquito Plant

Agastache cana

Mosquito plant is a tall wildflower with highly fragrant, very bright, pink flowers arranged in spikes. It is long-lived, and starts to bloom as early as late July and often continues into October. Both flowers and foliage bear a fragrance much like bubble gum. It is considered rare and endangered in New Mexico, where it is reported on only five sites in four counties. These sites are mountainous locations at about 5,000–6,000' (1,500–1,800 m). It also occurs in west Texas on a few sites.

J F M A M J J A S O N D

Mosquito plant is so easy to propagate from seed that it is surprising it is so rare in the wild. You can also start it easily from cuttings. It is easy to grow in most soils, though over-watering or high rainfall and high humidity are likely to cause problems. Even in semi-arid areas, it is not unusual to find a few seedlings in gardens.

White-lined sphinx moths and hummingbirds (both rufous and broad-tailed) are very frequent visitors. Mosquito plant is valuable for attracting hummingbirds to the garden over an extended period of time. Deer have little interest in browsing this plant. The common name refers to an alleged ability to repel mosquitoes; rubbing the leaves on skin or clothing is said to be quite effective.

rufous ♀ hummingbird

FIELD NOTES

■ Zones 5–8

■ Perennial

■ 3' (1 m)

☀ Full sun–part shade

◊ Moderate

Fireweed

Epilobium angustifolium

Fireweed, or willow–herb, is a tall wildflower with narrow leaves and showy, pink flowers in spikes. It is commonly found in disturbed areas along roads or in forest areas after either fire or logging. The flowers stand on a long, thin structure that develops into seedpods. In addition to seeds, fireweed spreads vigorously by root runners and can be invasive.

Fireweed will grow in most soils, but favors well-drained, moist situations. Propagate it from seed and from root shoots taken in spring.

Bees are attracted first to the female pollen-receptive flowers, at the bottom of the spike, which have the most nectar. They work their way up the flower spike until they encounter flowers in the male phase, where they pick up pollen before flying off to another plant in search of new pollen-receptive, nectar-rich flowers. These nectar-rich flowers make fireweed a valuable honey plant, and beekeepers have been known to search for them following forest fires or logging.

J F M A M J J A S O N D

bumblebee

fireweed seeds

FIELD NOTES
- Zones 2–9
- Perennial
- 4–5' (1.2–1.5 m)
- Full sun–full shade
- High

139

Gaillardia aristata hybrid

Blanket Flower

Gaillardia aristata

mason bee

Blanket flower is a showy wildflower, often with bright red and yellow flowers, common in grasslands of the western Great Plains. The name blanket flower is a reference to the similar colors and patterns in Southwestern weaving.

| J | F | M | A | M | J | J | A | S | O | N | D |

Blanket flower makes an attractive garden plant, and will bloom for a long time if the spent flowers are removed regularly. It will grow successfully in a wide range of soils, and will live longer if grown in moderately dry conditions. Propagate it from seed.

The color contrasts in these flowers serve to attract and direct pollinators such as bees and butterflies to the nectar and pollen as quickly as possible. Not all blanket flowers have both red and yellow petals, however. Some are all yellow, with only the center being red; these blanket flowers with all-yellow petals are the sole host of a moth that exactly matches the mottled reddish centers. It was named *Schinia masoni* for John T. Mason, a Denver butterfly collector, who discovered it in 1896. The color coordination is remarkable, and the moth lives only near Boulder, Colorado.

FIELD NOTES
- Zones 3–9
- Perennial
- 2–3' (60–90 cm)
- ☀ Full sun
- ◊ Low–moderate

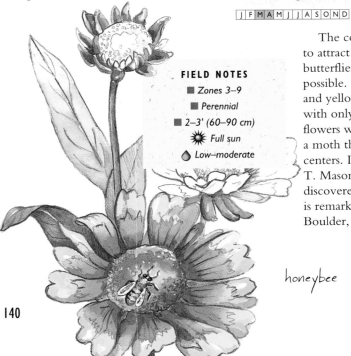

honeybee

140

Scarlet Gilia

Ipomopsis aggregata

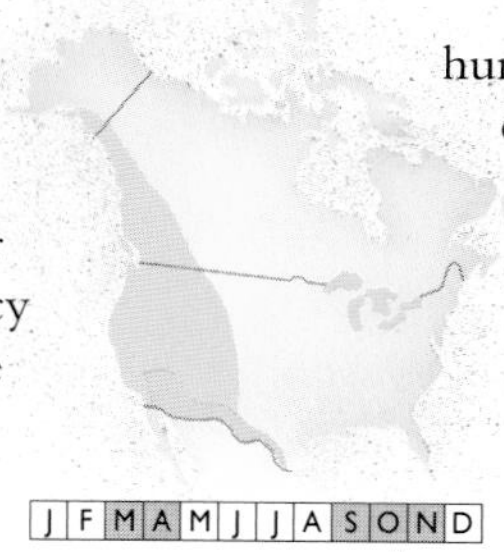

Scarlet gilia is a showy wild-flower with slender trumpets of red, pink, or white flowers along a tall flower stalk. The first year following germination, a clump of lacy leaves forms close to the ground. The second year, the plant produces the familiar, tall flower stalk and brilliant flowers. By the end of the second year, seeds have been dispersed and the entire plant dies.

Scarlet gilia favors well-drained soils. Propagation is usually from seed, sown in fall or early spring so it's likely to get sufficient winter weather to allow germination. Too much rainfall and too high summer temperatures are more limiting to good growth than minimum winter temperatures. Pinching back the flower stalks will stimulate rapid regrowth, resulting in more flowers and better-quality seed.

Scarlet gilia is a classic hummingbird- and sphinx-moth-pollinated flower. Both the birds and moths are well equipped to reach into the floral tubes. The petals bend back, making entry difficult for crawling insects. In this way, the flower reserves nectar for the moths or hummingbirds. Despite these defenses, bees sometimes rob the flowers by slitting the floral tubes near the nectar. Deer will browse scarlet gilia.

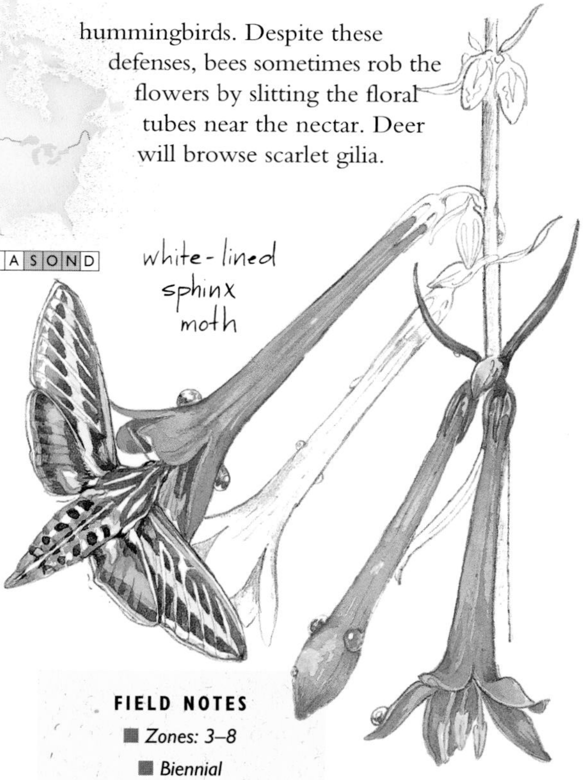

| J | F | M | A | M | J | J | A | S | O | N | D |

white-lined sphinx moth

FIELD NOTES

- Zones: 3–8
- Biennial
- 3' (1 m)
- Full sun
- Low–moderate

141

Western Blue Flag

Iris missouriensis

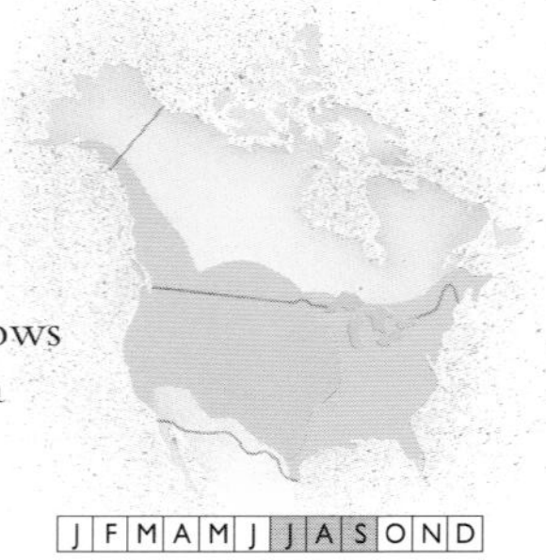

Western blue flag is strikingly beautiful when it blooms during May or June. It often forms dense populations of blue flowers in meadows that are wet in spring but dry later in the season.

Western blue flag will grow in a wide range of soils. You can propagate it from seed, which is very slow; propagation is quicker by dividing the roots after blooming in July. Though it prefers plenty of water in spring, this iris will tolerate much drier conditions in summer and fall.

| J | F | M | A | M | J | J | A | S | O | N | D |

bee leaving pollen

Bees pollinate these flowers by pushing their way between the long, upright style and sepal, or "fall", to reach the nectar near the center of the flower. In this process, they are brushed with pollen, while depositing pollen from previously visited flowers. Hummingbirds, usually considered pollination "good guys", are actually robbers of iris nectar. By probing between petals, sepals, and styles, they steal the nectar without accomplishing any pollination. Deer and most other browsers avoid all irises.

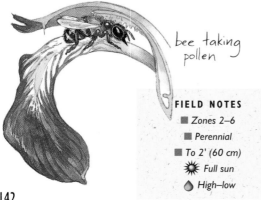

bee taking pollen

FIELD NOTES
- Zones 2–6
- Perennial
- To 2' (60 cm)
- ☀ Full sun
- 💧 High–low

142

Wild Four o'Clock

Mirabilis multiflora

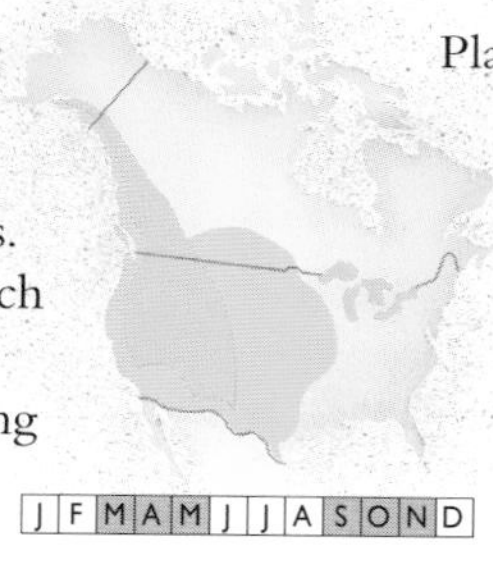

Wild, or Colorado, four o'clock is a wildflower of shrub-like proportions covered with numerous magenta flowers. There are often flowers throughout much of summer and fall. The flowers open during the late afternoon or early evening and close the following morning, unless the weather is cool and cloudy, in which case, they may remain open most of the day. This plant dies to the ground each winter.

The afternoon opening of wild four o'clock flowers offers gardeners a chance to create an interesting daily change of scenery. It has been called the immortal plant, a reference to its ability to return from the "dead" after rototilling or bulldozer activity. It will likely outlive the grandchildren of the gardener who plants it in a well-drained, sandy location.

Propagate it from seed or by digging small pieces of root in fall. Sow seed in spring or fall and transplant root pieces in fall. High rainfall and snowfall are more likely to limit good growth than minimum winter temperatures.

J F M A M J J A S O N D

Plants can develop and bloom fully with no summer rainfall, drawing on moisture stored in the roots.

Wild four o'clock is pollinated by dawn-to-dusk insects, mainly sphinx moths. Some plants are reported to produce a musky fragrance several hours after dusk, probably to attract more pollinating moths.

FIELD NOTES
- Zones 5–10
- Perennial
- To 3' (1 m)
- ☀ Full sun or light shade
- ◊ Low

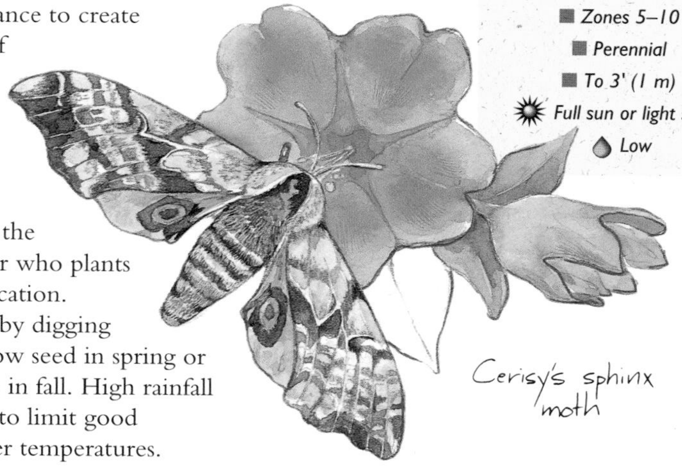

Cerisy's sphinx moth

Beardlip Penstemon

Penstemon barbatus

Beardlip penstemon, or scarlet bugler, is spectacular when its tall flower stalks are covered with bright red flowers. Its bloom time varies with altitude and from season to season: it usually blooms in mid-spring but sometimes in midsummer and in fall, too. All penstemons have five stamens (thus, the name penstemon), but the fifth stamen is often covered with hairs, resembling a bearded lip or tongue and giving rise to the names beardlip and beardtongue. The hairs actually help pick up pollen from bees as they push into the flowers looking for nectar.

| J | F | M | A | M | J | J | A | S | O | N | D |

Penstemons generally grow best in relatively dry, well-drained soil. Many are short-lived, but the life can be extended by clipping all but one or two flower stalks after blooming. A few seeds from the remaining stalks are enough to produce new plants in spring. In fact, seedlings often come up with no help from the gardener. Too much rain, snow, and humidity are more limiting to good growth than minimum winter temperatures.

Beardlip penstemon is mostly pollinated by hummingbirds. Bitter substances in some penstemons discourage browsing by deer.

broad-tailed
hummingbird

♂

FIELD NOTES
- Zones 2–9
- Perennial
- 3–4' (1–1.2 m)
- ☀ Full sun
- ◑ Low–moderate

144

Indian Pink

Silene laciniata

Indian pink, or Mexican campion, is a low-growing wildflower with delightful flowers of flaring, ragged, red petals. It is associated with partially shaded mountain slopes, frequently with pines and oaks, from west Texas to California.

This appealing little flower is easily grown in most soils that don't remain wet too long. It will grow in full sun in the North, but usually grows in part shade farther south. Indian pink is usually propagated from seed, which must be stratified to simulate winter conditions. Sow seed in fall or early spring. Too much rainfall and humidity in the eastern regions are probably more limiting to good growth than minimum winter temperatures.

Indian pink is a classic hummingbird-pollinated plant. Judging from the long period that hummingbirds remain at each flower, there must be a lot of nectar or it must be very good, or both. The flowers have backturned petals and a sticky section just below the petals that serve to repel crawling insects, who might otherwise steal the nectar without pollinating the flower.

| J | F | M | A | M | J | J | A | S | O | N | D |

red ants at base of flower

FIELD NOTES

■ Zones 4–8

■ Perennial

■ 1–2' (30–60 cm)

☀ Full sun–light shade

◯ Moderate

Scarlet Hedge Nettle

Stachys coccinea

At times, scarlet hedge nettle is covered with brilliant red flowers. It dies down during freezing weather but will continue growing in warm climates. Unlike its unpleasant namesake, it has no obnoxious stinging hairs.

Scarlet hedge nettle grows best in wet, nitrogen-rich soil, though it will survive remarkable drought by dying back and regrowing when conditions are wet again. Propagate it from cuttings or seed. Take cuttings when the plants are growing vigorously in late spring; sow seed indoors in pots and transplant seedlings in spring in cold climates, and at any time of the year in warm climates.

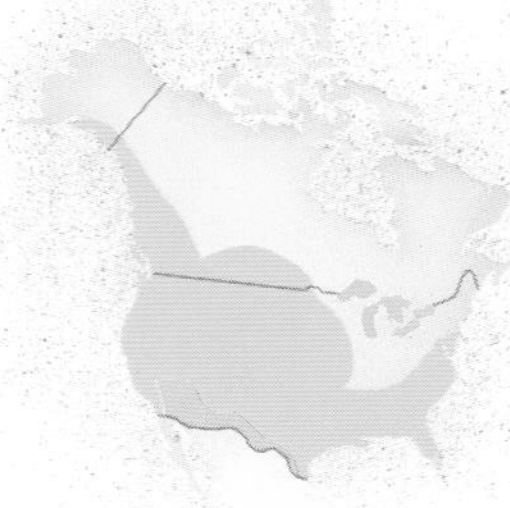

J F M A M J J A S O N D

Scarlet hedge nettle is likely to be the ultimate hummingbird plant: it seems to be absolutely irresistible, and it's not unusual to see humming-birds appear within minutes of setting a plant out on a patio. Because it will bloom well into the fall season, it is likely to be good for attracting hummingbirds in Flagstaff, Albuquerque, and Santa Fe, when the birds return from nesting in the central and northern Rockies.

painted lady butterfly

black-chinned hummingbird

FIELD NOTES
- ■ Zones 5–11
- ■ Perennial
- ■ Height: 1–3' (30–90 cm)
- ☀ Full sun–light shade
- ◊ High

Showy Goldeneye

Viguiera multiflora

Showy goldeneye is a pretty wildflower that is often found on disturbed areas in ponderosa pine (*Pinus ponderosa*) forests. Numerous flowers cover this plant for many weeks during summer, and the name goldeneye refers to the yellow flower center.

Showy goldeneye is easily grown in most soils. Extremely dry and extremely wet soils limit its growth, but it produces so many seedlings in moderate moisture conditions that it should be easy to establish with the grasses of midgrass prairies. Clipping off the seed heads will prolong the already-long blooming period. Propagation is easy from seed, which can be sown directly in the garden in spring.

Although the center of these flowers appears yellow to human eyes, bees and butterflies see reflected ultraviolet light. This probably creates a vivid contrast between the flower centers and the petals that is not visible to humans. To bees and butterflies, the primary pollinators of

showy goldeneye, this flower may look somewhat like blanket flower appears to humans. The pretty Gorgone crescentspot butterfly uses showy goldeneye as a host plant for its eggs, caterpillars, and chrysalises. Many other adult butterflies also visit this wildflower.

J F M A M J J A S O N D

FIELD NOTES

■ Zones 3–6

■ Perennial

■ 1–3' (30–90 cm)

☀ Full sun or part shade

◊ Moderate

Gorgone crescentspot butterfly

147

Saskatoon Serviceberry

Amelanchier alnifolia

Saskatoon serviceberry varies from a small, thicket-forming shrub to a small tree. The attractive, white flowers appear in early spring, followed by delicious berries. The fruit ripens earlier in the season than most wild fruit, giving rise to one of the common names, Juneberry. Other names include sarviceberry, shadbush, and shadblow. Shad refers to a fish that travels upstream at the time this shrub produces masses of flowers, or "blows".

Saskatoon serviceberry usually grows on moist sites; at higher elevations, it is more shrubby. It is adapted to a wide range of soils, and can be propagated from seed, which must be very well cleaned to eliminate germination inhibitors in the fruit. It may also be rooted from the many suckers that develop at the base of the plant.

Deer eat the twigs and leaves. Squirrels, chipmunks, and numerous birds eat the fruits with great enthusiasm. Grow serviceberry plants in cages if you would like to keep some berries for yourself! Coral hairstreak butterfly eggs overwinter, and the caterpillars feed on the developing fruits of serviceberries.

J F M A M J J A S O N D

spring flowers

FIELD NOTES
- Zones 2–7
- Deciduous
- 3–40' (1–12 m)
- ☀ Full sun–part shade
- 💧 Moderate

least chipmunk

148

Big Sagebrush
Artemisia tridentata

Extremely variable in size and shape, big, or basin, sagebrush is a pleasantly fragrant, silvery-colored shrub with three-lobed leaves. Clusters of tiny flowers are conspicuous but not showy. Blooming in August to October, bear in mind that they contribute significantly to hayfever.

Big sagebrush prefers well-drained soil, and seems to do poorly in landscaping if planted in clay soil. Do not overwater it, especially if the soil is not fast-draining. Propagate it from seed. Soil type and dry conditions are more important to good growth than minimum winter temperatures.

One of the interesting wildlife associations of this plant is the annual early spring appearance of tiny black aphids. These are followed in warmer weather by ladybugs, which eat the aphids, and finally, by ladybug larvae, which quickly finish off the remaining aphids. The whole cycle appears to be an example of leaving things alone to take care of themselves—no pesticide intervention is needed. Big sagebrush also provides protection and cover for rabbits and other small mammals.

| J | F | M | A | M | J | J | A | S | O | N | D |

ladybug beetle eating aphids

FIELD NOTES
■ Zones 4–10
■ Semi-evergreen
■ 1–8' (30–240 cm)
☀ Full sun–part shade
💧 Low

149

Chrysothamnus nauseosus

Rabbitbrush

Chrysothamnus spp.

Rabbitbrush is a small- to medium-sized shrub, with brilliant yellow fall flowers. The stems and leaves can be green, gray–green, or almost white. Sometimes, the seed heads are very showy well into winter. The genus name *Chrysothamnus* comes from the Greek words *chryso* for yellow and *thamnus* for shrub.

Rabbitbrush, with its showy flowers, attractive seed heads, and variously colorful winter twigs, is an important landscape plant for dry locations. Pruning back during winter results in a more compact plant with more flowers. Rabbitbrush will adapt to a wide range of soils, but overwatering is a problem in clay. Propagation is usually by seed. Too much rain, snow, and humidity are more limiting to good plant growth than minimum winter temperatures.

Blooming so late in the season makes rabbitbrush quite attractive to bees and butterflies, including the snout and monarch. The showy northern and sagebrush checkerspot butterflies use rabbitbrush as a host for their egg, caterpillar, chrysalis, and adult phases. Deer generally avoid browsing on this shrub. House finches and pine siskins eat the seeds.

J F M A M J J A S O N D

FIELD NOTES
- Zones 2–10
- Deciduous
- 2–6' (60–180 cm)
- ☀ Full sun
- 💧 Low

snout butterfly

150

Creeping Mahonia

Mahonia repens (Berberis repens)

Creeping mahonia is a low, spreading shrub, with holly-like, evergreen leaves and pretty, fragrant, yellow flowers from early to late spring. The summer and fall berries resemble grapes but have a unique flavor. Creeping mahonia is also known as creeping holly grape and Oregon grape, and there are many other similar references to holly and grapes. The leaves will change to a reddish color in fall if the plant is exposed to sun at that time of year. Native Americans used creeping mahonia to treat stomachaches, rheumatism, and, together with other plants, scorpion and spider bites.

J F M A M J J A S O N D

Creeping mahonia's chief value in the garden is as a useful groundcover that will grow in a variety of conditions. It is at its best in part sun and moderately moist situations.

spring flowers

When fully established, it tolerates considerable traffic and makes a good plant for edges of driveways and paths. It is adaptable to nearly all soil types. Propagate it from seed or by transplanting pieces of established plants before spring growth begins (this can be as early as February in Colorado). Plants grown in shade need to be moved into full sun gradually.

Fruit-eating birds will eat the berries, and the shrubs provide shelter for both birds and mammals. Deer seldom browse on this plant.

summer-fall berries

FIELD NOTES
- Zones 4–9
- Evergreen
- 1–3' (30–90 cm)
- ☀ Full sun–full shade.
- 💧 Low–high

151

Pinyon Pine

Pinus edulis

Pinyon pines are large, picturesque shrubs or small trees. The needles are very resistant to loss of moisture from the dry, prevailing winds. These pines also have both shallow and deep roots to make the maximum use of soil moisture. They produce a repellent odor and taste to discourage browsing of their valuable foliage.

The pinyon pine needs full sun and grows best on well-drained soil. You can grow the trees from seed but it is a slow process. Too much rainfall and humidity, and too cool summers, are more limiting to good growth than minimum winter temperatures.

Although pinyon pines can grow quite tall and wide, you can prune them to fit the garden landscape.

Birds and mammals take a large number of seeds and store them for eating later; but enough are not eaten after being stored in good growing places that the whole system continues. Clark's nutcrackers and Steller's and scrub jays all play a part in the seed distribution, but the pinyon jay is the primary partner in this seed distribution. Quail and wild turkeys also eat these seeds, as do various squirrels, voles, mice, bears, chipmunks, porcupines, and many human groups, both past and present.

J F M A M J J A S O N D

FIELD NOTES
- Zones 4–8
- Evergreen
- 10–40' (3–12 m)
- ☀ Full sun
- ◌ Moderate–low

pinyon ♂ jay

152

Gambel Oak

Quercus gambelii

Usually a large shrub, Gambel oak occasionally develops into a tree form and is most often found in sizeable thickets. The thickets are stimulated by fire and other disturbances such as logging. This results in vigorous, new trunks sprouting from the roots.

Gambel oak grows on a variety of soils but it favors well-drained soil. Propagate it from seed.

In the wild, the mammal and birdlife in Gambel oak shrublands is remarkable, and in gardens, creatures will be attracted to even a single Gambel oak. Scrub jays and rufous-sided towhees are especially conspicuous feeders on the acorns. In winter, scrub jays forage in the dead leaves and branches for nymph-filled galls, while mountain and black-capped chickadees search the branches for insects and eggs attached to the bark. About 50 species of mammal are associated with these oak woodlands, including mule deer, chipmunks, bears, rock squirrels,

J F M A M J J A S O N D

and many more. The Colorado hairstreak butterfly is especially associated with Gambel oak. It lays its eggs on the bark and the caterpillars feed on the leaves. Rocky Mountain duskywing and Horace's duskywing butterflies are also associated with this oak.

Colorado hairstreak butterfly

FIELD NOTES

■ Zones 3–7

■ Deciduous

■ To 35' (10.5 m)

☀ Full sun

◌ Moderate

153

Golden Currant

Ribes aureum

Golden, or buffalo, currant is a medium-sized shrub with pretty, golden-yellow spring flowers. The flowers have a delightful clove-like fragrance. The currants are black, purplish, or orange and ripen during summer. Shoshone Indians used golden currant roots to treat fevers, sore throats, angina, and dysentery. The fruits were an important food for the Plains Indians, either dried, eaten fresh, or made into jelly. Even the flowers are reported to be quite tasty.

Golden currant is adapted to a wide range of soils. Although it prefers moderate water, it is tolerant of dry to moist conditions. Propagate it from seed; grafted selections (cultivars) ensure fragrant flowers and tasty fruit.

A wide array of wildlife is attracted to the fruit, including skunks, squirrels, chipmunks, bears, and raccoons. Numerous birds often flock into the shrubs, finishing off the fruits in a feeding frenzy before the gardener considers them ripe enough to use. Tailed copper, gray comma, faunus anglewing, and zephyr anglewing butterflies all use currants as host plants for their egg, caterpillar, and chrysalis stages.

J F M A M J J A S O N D

summer fruits

zephyr anglewing butterfly

FIELD NOTES
- Zones 2–9
- Deciduous
- 2–6' (60–180 cm)
- Full sun–part shade
- Moderate

154

Datil

Yucca baccata

Datil, or banana yucca, is an agave-like shrub with machete-like leaves, which are much broader than those of most of the other yucca species. The fruit is almost banana-like. The shallow and deep roots, together with the thick leaves, serve to utilize any available soil moisture without losing much to the hot, dry conditions that prevail where this yucca grows. Native Americans in the Southwest used datil fiber cord to make sandals, mats, nets, baskets, and blankets; and roots to make a lather for washing. They ate the fruits raw or roasted and sun-dried, or used them to make a sweet drink. They also roasted and ate the seeds and flower buds.

Datil will grow in a wide range of soils, but good drainage is important since it won't tolerate wet soil. You can propagate it from seed, a slow method, or by offsets taken from established plants. Late fall or winter is the best time for this. Too much rainfall and humidity,

and too cool summers, are more limiting to good growth than minimum winter temperatures. Datil and other yuccas are pollinated by yucca moths, an interesting example of co-dependence, since neither can succeed without the other. Yuccas furnish protective nest sites for some birds and provide cover for numerous birds and small mammals.

| J | F | M | A | M | J | J | A | S | O | N | D |

brush mouse

FIELD NOTES
■ Zones 5–10
■ Evergreen
■ To 3' (1 m)
☀ Full sun
⬥ Low

155

Ponderosa Pine

Pinus ponderosa

Ponderosa pines are tall, majestic trees with long needles and fragrant bark, which smells like vanilla in warm weather. Their remarkable drought tolerance comes, in part, from their massive root system—at times, 35' (10.5 m) deep and 100' (30.5 m) across. Found over a broad section of the Western US, ponderosa pine is usually found above 5,600' (1,702 m) in the middle latitude of its range.

Ponderosa pines generally prefer well-drained soils. Transplant specimens in spring before the buds start growing—usually before early April. Although ponderosa pines will grow in a wide range of climates, too much rainfall is more limiting to good growth than minimum winter temperatures.

Abert's squirrels depend on fairly large areas of ponderosa pine forest. The little brown creeper hunts for insects in the thick, furrowed bark of these pines, and builds its nest behind pieces of loose bark. An estimated total of 57 mammals, including chipmunks, and 128 bird species, including the pygmy nuthatch, occur in this type of forest.

J F M A M J J A S O N D

male cone

FIELD NOTES
- Zones 3–8
- Evergreen
- 100–200' (30–60 m)
- Full sun
- Moderate

pygmy nuthatch

156

Narrowleaf Cottonwood

Populus angustifolia

Narrowleaf cottonwoods are tall, relatively narrow trees, and, as their name implies, they also have noticeably narrow leaves. They tend to grow in groves along mountain streams, usually between 3,200 and 9,900' (1,000 and 3,000 m).

All cottonwoods are considered phreatophytes (literally, "well plants"), meaning that they can survive only where the roots are able to draw on substantial amounts of subsoil moisture. The seeds of cottonwoods remain viable for only a few days, and must land on sunny, moist soil in this short time in order to germinate and grow successfully. Narrowleaf cottonwoods grow best in full sun and gravelly, well-drained soil, though they can adapt to other soils. You can propagate these trees by digging young sprouts from the roots of established trees. Transplant young trees before the leaves develop in early spring—usually, in March, April, or May, depending on altitude and climate. Beware—these trees sprout too vigorously from the roots for most urban situations.

Grouse and various songbirds, such as the black-headed grosbeak, eat the buds and young flowers. Beavers and other small mammals eat the buds, bark, and foliage. Mourning cloak and Compton tortoiseshell are among the many butterflies that use cottonwoods for their eggs, caterpillars, and chrysalises.

| J | F | M | A | M | J | J | A | S | O | N | D |

♂

black-headed grosbeak

FIELD NOTES
- ■ Zones 2–7
- ■ Deciduous
- ■ 60' (18 m)
- ☀ Full sun
- ⬤ High

157

Quaking Aspen

Populus tremuloides

Considered the most widely distributed North American tree species, quaking aspens are medium size and nearly always form beautiful groves. The bark is distinctively white, and the leaves, mounted on flattened leaf stalks, flutter in the slightest breeze (hence, the tree's scientific and common names). The fall color in the Rocky Mountains is a brilliant yellow.

J F M A M J J A S O N D

Aspens are adapted to a range of soils, as long as water is always available. Transplant specimens in early spring, before the leaves develop. Aspens prefer cool summers. In areas beyond their natural range, they suffer from many pests and diseases.

It has been estimated that 500 species of animal and plant, ranging from elk to fungi, depend on aspens. Grouse and quail browse on the buds, catkins, and seeds, and numerous cavity-nesting birds occupy old-growth aspen groves, including flammulated owls, northern saw-whet owls, northern pygmy owls, tree and violet-green swallows, house wrens, brown creepers, mountain bluebirds, pygmy nuthatches, and mountain chickadees. The red-naped sapsucker feeds on insects extracted from the bark and enjoys the sap. Tent caterpillars, aspen leaf-miners, and many other insects depend on aspens, and in turn, become the attraction for other species.

FIELD NOTES
- Zones 1–7
- Deciduous
- 40–60' (12–18 m)
- ☀ Full sun
- ◊ High

♂

red-naped
sapsucker

Riverbank Grape

Vitis riparia

Riverbank grape is widespread in the Mountains and Basins and Desert Southwest regions. It tends to be found near water, usually streams, and can ramble over large shrubs and small trees. It is relatively late to leaf out in spring and suffers seriously from late-spring freezes, which can eliminate the grape crops—a significant loss for wildlife. Pueblo Indians cultivated this vine for the fruits, which they ate raw or dried in the sun for eating later. The grapes make a tasty jelly, juice, and wine. The tendrils can be eaten raw as a different and distinctive tidbit, and the leaves are reported to quench thirst.

Riverbank grape is adapted to a range of soils. Propagation is usually by seed, which must be totally cleaned of germination-inhibiting chemicals in the fruit. Riverbank grape will grow well over a trellis, arbor, or fence in the garden.

J F M A M J J A S O N D

The wildlife value of this plant is immense: the grapes provide food for numerous mammals, including squirrels and raccoons, and possibly as many as 100 species of bird. In addition to food, the vines provide good cover for nests.

raccoon

FIELD NOTES
- Zones 3–8
- Deciduous
- To 20'+ (6 m+)
- ☀ Full sun
- ◐ High

159

Pale evening primrose (left); Rocky Mountain bee plant (right)

Additional species

	Plant name/flower color	Zone	Height	Visitors/attraction/season
CACTI	Porcupine prickly pear (yellow/red) *Opuntia erinacea*	5–8	6" (15 cm)	bees/flowers/spring, summer
HERBACEOUS FLOWERS	Rocky Mountain columbine (blue) *Aquilegia caerulea*	3–8	1–2' (30–60 cm)	hummingbirds/flowers/summer
	Rocky Mountain bee plant (pink/white) *Cleome serrulata*	5–9	2–4' (60–120 cm)	bees/flowers/summer, fall
	Small-leaf geranium (pink) *Geranium caespitosum*	4–8	2'± (60 cm±)	bees/flowers/spring–fall
	Sweet-flowered four o'clock (white) *Mirabilis longiflora*	7–10	1–3' (30–90 cm)	moths/flowers/summer, fall
	Bee balm (white) *Monarda citriodora*	5–9	1–3' (30–90 cm)	bees/flowers/spring–summer
	Pale evening primrose (white) *Oenothera pallida*	4–7	6–12" (15–30 cm)	moths/flowers/spring–fall
	Desert penstemon (pink) *Penstemon pseudospectabilis*	5–8	4–6' (120–180 cm)	bees/flowers/spring, summer
	Rocky Mountain penstemon (blue) *Penstemon strictus*	4–9	2–4' (60–120 cm)	bees/flowers/spring, summer
	Wandbloom penstemon (pink) *Penstemon virgatus*	4–7	2–4' (60–120 cm)	caterpillars/leaves/spring–fall
	Thread-leaf groundsel (yellow) *Senecio douglasii* var. *longilobus*	5–9	1–3' (30–90 cm)	bees, butterflies/flowers/summer, fall
	Fendler's globemallow (pink) *Sphaeralcea fendleri*	4–9	2–5' (60–150 cm)	bees/flowers/summer, fall
SHRUBS	Fernbush (white) *Chamaebatiaria millefolium*	5–8	2½–5' (75–150 cm)	bees/flowers/summer
	Cliff rose (cream) *Cowania mexicana*	5–9	4–20' (1.2–6 m)	bees/flowers/spring–fall
	Feather dalea (purple) *Dalea formosa*	5–9	2' (60 cm)	bees, butterflies/flowers/spring–fall
	New Mexico privet (n/a) *Forestiera neomexicana*	5–9	6–10' (1.8–3 m)	birds/seeds/fall
TREES	Utah serviceberry (white) *Amelanchier utahensis*	4–8	25' (7.5 m)	birds, mammals/fruits/fall
	Engelmann spruce (n/a) *Picea engelmannii*	3–9	80–115' (24–35 m)	chickadees/cones/variable
	New Mexico locust (pink) *Robinia neomexicana*	6–8	25'± (7.5 m)±	porcupines, chipmunks, Gambel's quail/seeds/fall
VINE	Arizona honeysuckle (red) *Lonicera arizonica*	6–9	variable	mammals/seeds/fall hummingbirds/flowers/summer

The Desert Southwest

THE DESERT SOUTHWEST
The Southwestern Garden

The warm deserts of the Southwestern United States are found from 5,000' (1,500 m), where desert meets juniper–oak woodlands, to below sea level, in the Imperial Valley.

Unlike the colder deserts farther north, moisture is measured not in snow but in rainfall, which averages 4–8" (100–200 mm) a year. The two main deserts, the Chihuahuan and the Sonoran, are mostly in Mexico but cross the border into the United States. The Chihuahuan Desert is in southwestern Texas, southern New Mexico, and southeastern Arizona, and the Sonoran Desert is in Arizona, southern Nevada, and southeastern California. The Sonoran Desert is divided into a colder portion, parts of which are called the Mojave Desert, and an almost frost-free portion, called the Lower Colorado Desert.

Each desert has signature plants and a special character of its own: the Sonoran Desert has saguaro (*Carnegiea gigantea*) and the Mojave Desert has Joshua tree (*Yucca brevifolia*). All warm deserts have desert willow (*Chilopsis linearis*), ocotillo (*Fouquieria splendens*), mesquites (*Prosopis* spp.), creosote bush (*Larrea* spp.), and some form of globe-mallow (*Sphaeralcea* spp.). The prettiest portions of the deserts are in the foothills, where higher elevations mean relatively cooler temperatures.

A natural desert garden is composed of small, native trees such as the desert willow and the blue palo verde (*Cercidium floridum*); flowering and ever-green shrubs, such as daleas and fairydusters (*Calliandra* spp.); succulent accents such as cacti (*Opuntia, Carnegiea* and *Ferocactus* spp.), yucca, or agave; and a myriad of small, rainy season flowers and grasses.

A drip-irrigation system will prolong bloom times and keep non-indigenous trees and cacti alive. Birds favor cacti for nesting sites, because the spines discourage predators and the pads provide food and moisture. While the plants are very small, wire cages will protect them from hungry rabbits.

It is very easy to make a Southwestern garden attractive to wildlife, because chuck-wallas and other lizards, bejeweled beetles, ground squirrels, hummingbirds, quail, butterflies, and hawk moths abound in the desert. A water faucet, within sight of the house and on a timer to drip for an hour at dawn and again at sunset, will bring all the local desert creatures to your Southwestern garden.

Saguaro

Carnegiea gigantea (Cereus giganteus)

For many people, the saguaro symbolizes the Southwest, yet this tree-like cactus is native only to the Sonoran Desert. Several arms branch from a central trunk, all fleshy and succulent with stored water. The spring flowers, which appear at the tips, are white and are said to smell of ripe melon. The fruits, juicy and sweet/tart, ripen in June.

Saguaros are not easy to propagate, but a few nurseries are now attempting to produce 1–3' (30–90 cm) high plants at a reasonable price. Transplanted saguaros have a 50 percent mortality rate. Good drainage is essential, so never plant saguaro in a swale. Drip irrigation is important for the first year, while the plant gets established.

J F M A M J J A S O N D

The flowers are pollinated at night by nectar-sipping lesser long-nosed bats and long-tongued bats. Day visitors include white-winged doves, butterflies, and hummingbirds. The fruits are eaten by bats, white-winged doves, songbirds, humans, foxes, ringtails, and other mammals. Woodpeckers and carpenter birds use their beaks to carve homes in the trunks. Birds of prey build nests in the branches. Fallen logs house lizards and beetles.

lesser long-nosed bat

FIELD NOTES
- Zone 9
- *Native to Sonoran (frost sensitive and limited)*
- *Evergreen*
- *20–50' (6–15 m)*
- ☀ *Full sun*
- ◌ *Low*

Claretcup Cactus

Echinocereus triglochidiatus

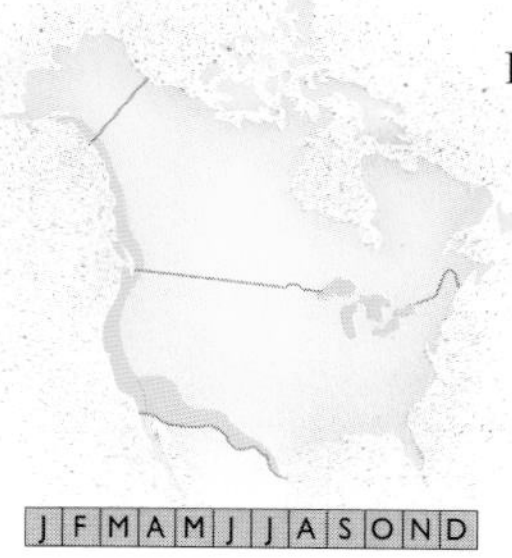

Claretcup cactus, or Mojave mound, flowers may be brilliant orange, red, or magenta. Minerals in the soil rather than genes may determine the exact shade of red. The blooms are unexpectedly large for such a small plant and very profuse. They appear in early spring to early summer and are followed by 1" (2.5 cm) juicy, red fruits. The cylindrical cactus stems increase in number with age, so the plant gets wider, not taller. A tiny, nursery-propagated claretcup cactus may be 18" (45 cm) wide in a few years and ultimately 4' (1.2 m) wide.

J F M A M J J A S O N D

Propagation is by seed, tissue culture, or division. Large specimens for sale have usually been poached, so ask to see a state or federal permit and check with your county agricultural extension agent before buying large plants. Cacti require good drainage at all times. On desert flats, they need some watering. If the stems turn brown at the base, it is a sure sign of too much water—the most common killer of cacti.

The flowers are a source of nectar for Costa's hummingbirds, Scott's orioles, and butterflies. Songbirds, white-winged doves, and many species of desert mammal, eat the fruits.

FIELD NOTES

- Zone 9
- Adapted to southern WC
- Evergreen
- Less than 1' (30 cm)
- ☀ Full sun to a little shade
- ◌ Low

Costa's hummingbird

165

Teddy-bear cactus (*Opuntia bigelovii*)

Prickly Pear, Cholla

Opuntia spp.

The stems of prickly pear are round or elliptical flattened pads. The silhouette is dense. Chollas' stems have tubular joints and are much taller and airier in structure. Flowers for both are large and gaudy—usually yellow, orange, hot pink, or magenta, but occasionally a pale yellow-green. These appear in spring or summer. The fruits ripen in late summer or fall. They are juicy, sometimes spiny, and most turn red, mauve, or yellow.

J F M A M J J A S O N D

Propagation is simple: remove a pad or joint, allow the wound to harden and dry, and then place the pad or joint on dry soil. Cacti are very sensitive to wet or even moist soil, which can turn them brown at the base and then cause them to rot. A cactus needs water only if it shrivels and wrinkles. Place these cacti away from paths and patios to avoid contact with the painful spines. Chollas must be planted with great care, as the joints readily detach.

Bees visit the pollen-rich flowers, and desert mammals and birds eat the fruit. The prickly branches are used for nesting by cactus wrens and road runners. In drought, many animals eat the pads for their water and food content.

prickly pear flower

cactus wren

FIELD NOTES

■ Zones 4-10

■ Some species are found over most of north, central, and south America; most native to Desert Southwest

■ Evergreen

■ 1–15' (30 cm–4.5 m)

☀ Full sun to a little shade

⬗ Low

166

Desert Marigold

Baileya multiradiata

The low-growing, pale blue-green foliage of desert marigold looks handsome in all seasons. The daisies, which top slender, leafless stems, appear year-round, except when heat or cold are at their fiercest. The yellow flowers are multilayered with matching yellow centers.

The tiny seeds ripen and scatter over a period of several months. Although desert marigold lives only one to two years, you can maintain a colony by allowing it to self-sow. Let these flowers weave themselves in and out of your other plantings: the garden will look a little different each year, but it will always be attractive. Good drainage is essential.

J F M A M J J A S O N D

A mulch of decomposed granite or a natural desert covering of rocks hides enough of the seed from predators to ensure a continuing crop of plants.

The nectar-rich flowers attract butterflies, such as the Leanira checkerspot and the desert and pima orangetips. Birds, insects, and small mammals eat the seeds. Rabbits are fond of the foliage, so protect young plants with wire cages until you have enough desert marigold to share. The desert grassland whiptail lizard may be found sheltering beneath the foliage. Horses also like the foliage, but it seems to be poisonous to other domestic animals.

desert grassland whiptail lizard

FIELD NOTES

- ■ Zones 8–10
- ■ Annual or short-lived perennial
- ■ Evergreen
- ■ 12–18" (30–45 cm)
- ☀ Full sun to a little shade
- 💧 Low–moderate

167

Parry's Penstemon

Penstemon parryi

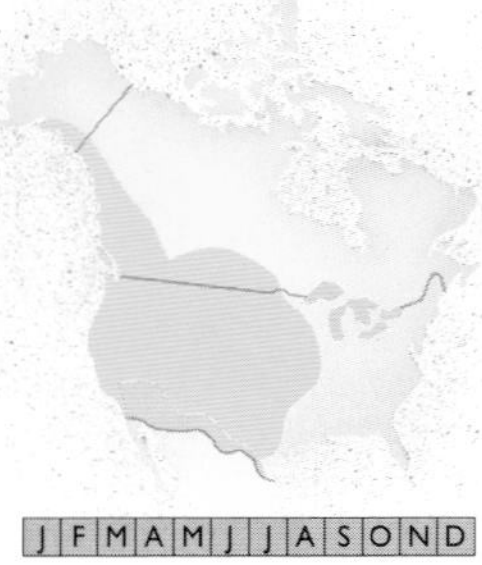

Every spring, several stems emerge from a single rosette to bear clusters of tubular, rosy pink flowers. Small, round capsules within the flower head quickly ripen to tan and release tiny, black seeds. The stems then die back, and only the rosette remains green all year. There are many attractive species of penstemon, ranging in color from blue and lavender to white.

J F M A M J J A S O N D

Propagation is easy from seed, and most gardeners let their penstemons self-sow. Start with at least three penstemons of one species to assure good cross-pollination. Mulch them with de-composed granite: the granite hides the tiny seeds from birds and insects, and they can wait for rain to trigger germination. Ensure there is constant good drainage and water sparingly—the result should be a bumper crop every spring.

Large clumps of penstemon several years old will have an almost constant stream of hummingbirds darting in and hovering while probing the flowers for nectar with their beaks. The leaves host butterfly larvae and are browsed by deer.

FIELD NOTES
- Zones 8–10
- Perennial
- Evergreen rosette
- 3–4' (1–1.2 m)
- Full sun to a little shade
- High

♂

blue-throated hummingbird

New Mexico Agave

Agave neomexicana

rosette base

New Mexico agave looks like a large, blue-green artichoke with beautiful but dangerous black-tipped spines. It is much smaller than the giant maguey (*A. americana*), making it a more suitable size for home landscapes. The flower stalks look like a candelabrum of golden yellow scrub brushes. These plants flower once only, in spring, when 8–20 years old. After setting seed, the original plant dies, but baby agaves, called offsets or pups, then appear at the base.

These pups will form a colony, or they can be cut free with a clean knife, allowed to air-dry, and then planted in a new spot. Soil should be open and gravelly with low organic content, and always well drained.

J F M A M J J A S O N D

Agaves are very easy to grow, but they rot if they are overwatered. The flowers are a source of nectar for butterflies, hummingbirds, lesser long-nosed bats, and long-tongued bats. The leaves serve as food for the larvae of the agave borer butterfly. The Lucifer hummingbird nests in the rosettes, and the agave woodpecker nests on the dead flower stalks.

long-tongued bat

FIELD NOTES
- Zones 7–10
- Evergreen
- 2' (60 cm) with much taller bloom stalk
- Full sun to a little shade
- Moderate

Ajamete

Asclepias subulata

Ajamete, or rush milkweed, is one of the most drought tolerant of the milkweeds. It has a grassy texture composed of a dense cluster of very pale green stems. The flowers are small and creamy white, usually appearing in March. The seedpods, like other milkweeds, are large and filled with fluffs of silvery silk.

Propagation is easiest by root cuttings or division, but seed may be sown any time there might be at least two months without either frost or scorching drought. Ajamete grows in deep soil along washes, and stays green in summer only when moisture is always present. Although it likes water, it must always be well drained.

Ajamete is important for monarch butterflies. They lay their eggs on the stems and their larvae feed on the plant.

The butterflies return to sip nectar from the flowers, where they are often seen alongside the tarantula hawk wasp.

J F M A M J J A S O N D

milkweed seedpod

monarch butterfly

FIELD NOTES
■ Zones 9-10
■ Evergreen or drought dormant
■ 3–5' (1–1.5 m)
☀ Full sun to a little shade
💧 Low–moderate

170

Red Barberry

Berberis haematocarpa (Mahonia haematocarpa)

Red barberry is a large shrub with tiny, dark gray-green leaves that are as prickly as holly. In very early spring, red barberry is fragrant with a profusion of small, golden yellow flowers. These are followed in early summer by red berries. Frosts turn the foliage purple, the color lasting throughout winter.

Collect the fruit when ripe, clean and air-dry the seed, and store the seed in a tightly sealed container in the refrigerator. In fall, or after the last frost in spring, plant the seed ¼" (5 mm) deep directly in the ground where you want it to grow;

J F M A M J J A S O N D

transplanting is hard, because the root system develops extensively before top growth begins. Cuttings are difficult without professional equipment, such as a mist bench. Once established, red barberry is long-lived if its soil is well drained and more dry than moist.

Nectar-seeking butterflies and bees visit the flowers, and mammals and birds, such as white-winged doves, verdins, and desert sparrows, eat the sweet fruits. The spiny, dense branches provide cover for many small desert creatures. The roots and bark have medicinal properties and the stem yields a yellow dye.

black-throated sparrow ♂

FIELD NOTES
- Zones 7–9
- Native to Pinyon/Juniper belt of Desert Southwest
- Evergreen
- 6–12' (2–4 m)
- ☀ Full sun to a little shade
- ◊ High

171

Woolly Butterflybush

Buddleia marrubbiifolia (Buddleja marrubbiifolia)

The main attraction of woolly butterflybush is its velvety soft leaves of pale, silvery gray. The flowers are not showy from a distance, but when seen close up, they are delightful balls of tiny orange blossoms. Flowering is rarely profuse; small quantities of flowers appear after rain during the warmer months, thereby producing nectar over a long period of time.

Propagation is easy from untreated seed in warm soil. The seed is ripe in mid- to late summer. Keep the seed dry and well ventilated until it is planted in early fall or late spring. Woolly butterflybush is easily damaged by cold in El Paso and even in Las Vegas. A little extra water in spring and fall increases the number of flowers, but excellent drainage is necessary at all times.

The flowers are a source of nectar for many species of butterfly, including lyside, Milbert's tortoiseshell, Empress Leilia, and Vesta and Texan crescentspots, as well as hummingbirds and bees. Sheep, goats, and deer browse the foliage.

velvety leaves

J F M A M J J A S O N D

Milbert's tortoiseshell butterfly

FIELD NOTES

- Zones 9–10
- Evergreen
- 3–5' (1–1.5 m)
- ☀ Full sun to a little shade
- ◗ Moderate

Pink Fairyduster

Calliandra eriophylla

After rain, pink fairyduster, or mesquitillo, covers itself with flowers. These unusual blossoms look like plumy 2" (5 cm) featherdusters. Baja fairyduster (*C. californica*) is red. With a small amount of extra water, a fairyduster's flowers completely obscure its tiny green leaflets. If severe drought or frost follows, the flowers and leaves drop off, revealing the short, intricately branched shrub. In a garden with drip irrigation, this plant usually remains evergreen. Narrow, 1–3" (2.5–7.5 cm) velvety beanpods ripen in summer.

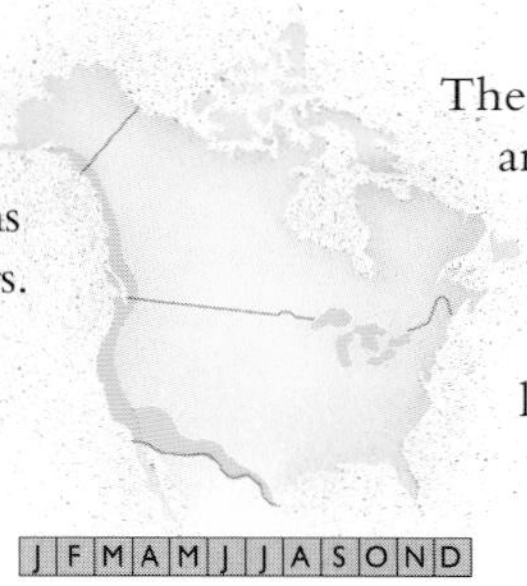

Both kinds of fairyduster are available in nurseries. Growing your own is fairly easy from root cuttings or seed. Fairyduster requires well-drained soil. Low growing and easily maintained below knee height, it is often used as a soft, nonspiny desert groundcover. It spreads by the roots to form a small thicket, and is useful for controlling soil erosion.

One easy way to keep fairyduster short is to have deer nearby; they love to browse it and fairyduster seems to benefit from the attention.

The seed is a favorite of bobwhites and quail. The flowers are a source of nectar for desert swallowtail, short-tailed black swallowtail, desert orangetip, and painted lady butterflies, as well as for bees.

J F M A M J J A S O N D

short-tailed black swallowtail butterfly

FIELD NOTES

■ Zone 9

■ Evergreen, drought deciduous, or cold deciduous

■ 2–4' (60–120 cm)

☀ Full sun

💧 Moderate

173

Bush Dalea

Dalea pulchra

There are many lovely daleas in the Southwest that are short, shrubby, and covered with purple flowers at least once during the year. Bush dalea is native to southern Arizona. Its leaves are small and silvery, and its flowers are a rich blue-purple and are most likely to appear in early spring. The seeds are small and not very noticeable. Feather dalea (*D. formosa*) is cold tolerant to Zone 7. It has gray-green leaves and feathery, plumed seeds in addition to rose-purple flowers that might appear in spring or fall, or both. Gregg dalea

J F M A M J J A S O N D

(*D. greggii*), winter hardy to Zone 8, is so low growing that it is often used as a groundcover.

Propagation is by fresh, untreated seed sown in early spring or by semi-hardwood cuttings. These shrubby daleas love limestone and rocky soil with low organic content, but they can adjust to richer conditions as long as drainage remains excellent.

The nectar of the flowers is very sweet and attracts both butterflies and several species of bee, including the bumblebee. Often, a bush dalea is heard before it is seen, because of the melodious buzz of busy nectar-eating insects. Chewed leaves might indicate the presence of dogface butterfly larvae. The spotted ground squirrel feeds on dropped dalea seeds.

spotted ground squirrel

FIELD NOTES

■ *Zone 9*

■ *Evergreen, cold deciduous, or drought deciduous*

■ *3–5' (1–1.5 m)*

☀ *Full sun*

◊ *Moderate*

174

Brittlebush

Encelia farinosa

In late winter and early spring, brittlebush, or incienso, bursts into bloom with hundreds of yellow daisies. Where native, it is so showy and ubiquitous that spring would be hard to imagine without it. A medium-sized plant, the gray leaves form a compact mound topped by bare stems, in turn topped by a myriad of flowers. Summer drought renders the silvery stems bare and leafless. With irrigation, they remain leafy but get leggy. Most people cut back the stems to the ground in late spring. New leaves—fuzzy, silver, and aromatic—appear rapidly. You can pull brittlebush out by the roots if there are enough seedlings to replace it.

Seedlings are usually abundant because brittlebush self-sows. Fresh, untreated seed germinates after rain in spring and fall. This shrubby flower remains short and dense if it receives intense reflected heat and gets little or no extra water. Too much water makes it overlarge and floppy, while too much shade or winter cold retards its flowering.

The blossoms are very popular with butterflies and bees. The chuckwalla loves to browse brittlebush, while songbirds and deer mice eat the seeds. The dried branches can be burned as incense, giving rise to the common name incienso.

| J | F | M | A | M | J | J | A | S | O | N | D |

deer mouse

FIELD NOTES

■ Zones 9–10

■ Adapted to southern WC

■ Evergreen or drought deciduous

■ 2–6' (60–180 cm)

☀ Full sun

◯ Low–moderate

175

Ocotillo

Fouquieria splendens

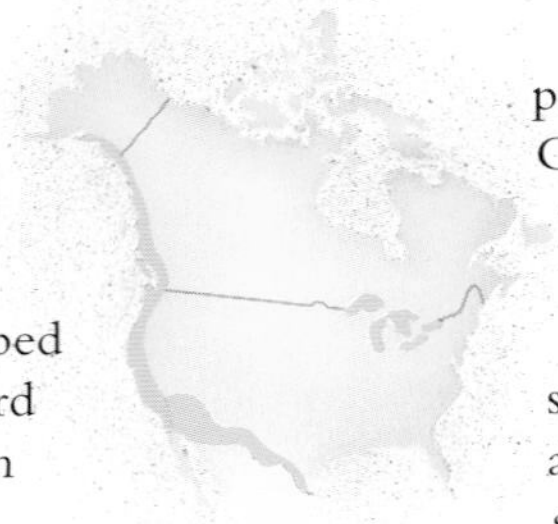

It is useful for the desert landscape designer to think of ocotillo as a tree because it is so tall. Instead of a trunk or spreading branches, however, it has a vase-shaped cluster of spiny stems, tapering inward toward their roots. The stems remain green with occasional irrigation. After every warm-season rain, the ends of the stems become bright with magnificent, large, red to orange flowers.

The easiest way to propagate ocotillo is from cuttings taken any time of year, but it is also easy to propagate from seed. Collect the seed in summer as the capsules turn brown, but before they split. Dry the seed thoroughly and store it in the refrigerator. Sow the seed in fall or spring, barely covering it with soil, and water very lightly. Too much moisture for cuttings, seedlings, or the mature plants, causes rotting and fungal diseases. Good drainage is essential.

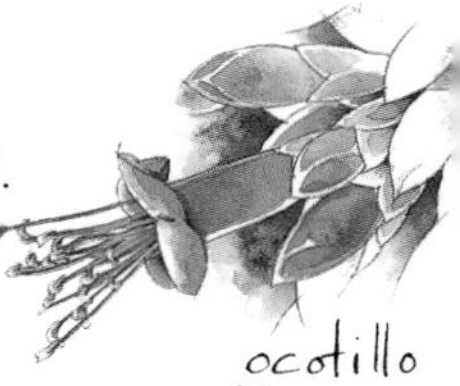

The flowers are a source of nectar for hummingbirds, orioles, butterflies, and bees. Small songbirds build nests on the spiny stems.

J F M A M J J A S O N D

ocotillo flower

FIELD NOTES
■ Zones 8–10
■ Adapted to southern WC
■ Stems are dormant in drought; leaves appear only after rain
■ 12–25' (4–7.5 m)
☀ Full sun
💧 Low

hooded oriole ♂

176

Chuparosa

Justicia californica (Beloperone californica)

Chuparosa is a sprawling shrub with evergreen stems so dense that the absence of leaves most of the year is hardly noticeable. Leaves actually do appear, but only for a short time after heavy rain. Chuparosa has no spines, making it safe to plant by paths and doorways. It is covered with hundreds of small, red flowers in March to May, and scattered blooms continue to appear throughout summer and fall.

Propagation is best from softwood cuttings. Seeds are rarely used because they are so hard to collect at the right time; they pop as soon as they are ripe. Growing chuparosa in a garden is easy if drainage is excellent and irrigation is sparse. It is commonly used as a flower or evergreen shrub (not hedged, of course, because of the leafless green stems), but it can also climb up a palo verde tree or a trellis, and it drapes itself becomingly over a wall or the sides of a large patio pot.

Several species of hummingbird, and butterflies, orioles, warblers, and goldfinches visit the flowers. Quail and house finches eat the tiny seeds. Chuparosa is also a larval plant for the Chara checkerspot butterfly.

J F M A M J J A S O N D

Calliope
hummingbird

FIELD NOTES
■ Zones 9–10
■ Adapted to southern WC
■ Evergreen stems
■ 5–8' (1.5–2.4 m)
☀ Full sun to dappled shade
◊ Moderate

177

Texas Ranger

Leucophyllum frutescens

Texas ranger and the other cenizos are all attractive, low-growing, silver-leaved, velvet-textured shrubs native to the Chihuahuan Desert. Texas ranger has the most cold tolerance, but boquillas silverleaf (*L. candidum*) and Sierra cenizo (*L. revolutum*) have the palest, showiest leaves.

J F M A M J J A S O N D

The blossoms come in colors ranging from deep purple-blue to pink or pure white. These flowers appear after rain in spring, summer, and fall and are sometimes so numerous that the silvery leaves are almost hidden. The fruits are tiny and tan-colored.

Propagation is easy by seed or cuttings, but there are so many pretty selections for sale that most people buy their cenizos. In the garden, give cenizos plenty of sun and good drainage. Pruning keeps them small, but old specimens with twisting silver trunks are also attractive.

Several species of butterfly and bee visit the flowers. Theona checkerspot, a pretty orange-and-black butterfly, uses cenizos as its larval plants as well as for nectar. The larvae are velvety, very dark brown, dotted and banded with cream, and decorated with many branching spines.

FIELD NOTES

- ■ Zones 8–10
- ■ Native to Texas (Chihuahuan Desert)
- ■ Evergreen
- ■ 4–8' (1.2–2.4 m)
- ☀ Full sun
- ◊ Moderate

Theona checkerspot butterfly

Littleleaf Sumac

Rhus microphylla

Littleleaf sumac is a large shrub with small, dark, shiny green leaves. Occasionally, it forms a multitrunked tree. The clusters of white flowers are quite profuse in early spring. The bright orange-red fruits appear in early summer and last for several weeks. Littleleaf sumac is evergreen during mild winters, but whenever frosts make fall color possible, it turns beautiful shades of red, orange, and purple.

summer fruit

Littleleaf sumac can be challenging to grow from seed. The best way is to pick seed that is almost ripe but still soft, and to plant it immediately ½" (1 cm) deep exactly where the plant is to grow. Place a wire cage around the spot and keep checking it for several months. If it doesn't germinate in fall, it is most likely to appear in early spring after all danger of frost is past. Good drainage is necessary, but any soil is acceptable, even gypsum.

Butterflies and lots of bees visit the flowers. Gambel's quail, chipmunks, and other creatures eat the fruits. Mule and other deer browse on the leaves.

| J | F | M | A | M | J | J | A | S | O | N | D |

Gambel's quail

FIELD NOTES
- Zones 8–9
- Evergreen, drought deciduous, or cold deciduous
- 8–25' (2.4–7.5 m)
- ☀ Full sun to a little shade
- 💧 Moderate–high

Dorri Sage

Salvia dorrii

There are many salvias (the true culinary sages) native to the Desert Southwest. They come in shades of white, red, pink, and purple. All of them are pretty and aromatic, and most of them are shrubby. Dorri, or desert, sage is easily the most stunning. It has pale, blue-gray leaves that stay evergreen in mild winters. The flowers are electric-blue petals with yellow stamens nestled into fluffy balls of reddish purple.

Propagation is by softwood cuttings or fresh seed sown in spring. Dorri sage prefers a sandy, loose, well-drained soil with a low organic content. Another sage easily found in Southwestern nurseries is autumn sage (*S. greggii*).

J F M A M J J A S O N D

It is not so drought tolerant and requires weekly irrigation.

Flowers of dorri sage are deep-throated enough to be used by hummingbirds, and shallow-throated enough to be visited by butterflies and bees. The aromatic leaves are shunned by all but the hungriest rabbits and deer. Their pungency changes with the seasons, but when they are flavorful and not bitter, they can be used in cooking and to make herbal tea.

honeybees

FIELD NOTES
■ *Zones 8–10*
■ *Evergreen or cold deciduous.*
■ *2–3' (60–90 cm)*
☀ *Full to a half day of sun*
◗ *Moderate–high*

180

Globemallow

Sphaeralcea ambigua

Globemallow, or apricot mallow, is a small shrub covered with cup–like flowers in spring. The 1" (2.5 cm) blooms are translucent and glow in the sun. The most common color is orange, ranging from pale peach to almost red, but other shades are pink, white, and lavender. The leaves are light green and are coated with silvery fuzz. In an irrigated garden, these leaves will be evergreen.

Propagation is easy from fresh, untreated seed. Most gardeners let their globemallows self-sow. Good drainage is important; and sandy soil, with a mulch of decomposed granite provides the perfect setting to keep a colony going for several years. If the foliage looks untidy after flowering, cut the shrub back to 4" (10 cm) and let fresh, new foliage grow before temperatures get too hot.

Butterflies and bees visit the flowers, and lizards shelter beneath the shrub. The leaves are browsed by deer and eaten by the larvae of several species of butterfly, including the common checkered skipper, the streaky skipper, and the painted lady. The desert spiny lizard chases insects attracted to the flowers and foliage.

J F M A M J J A S O N D

spring flower

desert spiny lizard

FIELD NOTES

■ Zones 8–10

■ Evergreen or drought deciduous

■ 2–3' (60–90 cm)

☀ Full sun to a little shade

⬥ Moderate

Roemer Acacia

Acacia roemeriana

Roemer acacia is a small tree with feathery leaves and white, fluffy balls of extremely fragrant flowers. Unlike most acacias, its stems and branches are not very prickly. An old tree might have a trunk 12–16" (30–40 cm) in diameter, but multitrunks of 3–4" (8–10 cm) in diameter are more common. The flowers appear in spring, except in drought years, when they occur following the first good rain. Seedpods are broad and flat and turn a warm pink-tan when ripe.

Plant seed in spring. If germination is not rapid, scarify the seed and try again. Softwood cuttings may be taken in late spring. Roemer acacia prefers deep soil with good drainage. It grows quickly but rarely lives more than 50 years.

J F M A M J J A S O N D

There are many Southwestern acacias and they all provide nectar for butterflies, such as the marine blue, and pollen for bees. Birds, such as the broad-billed hummingbird and the black-tailed gnatcatcher, nest in the branches. Quail eat the seeds, which can be ground into an edible meal called pinole. Deer browse on the leaves and pods.

seedpod

FIELD NOTES
■ Zones 8–10
■ Deciduous in winter
■ 10–25' (3–7.5 m) tall with broader width
☀ Full sun
💧 High

marine blue butterfly

Blue Palo Verde

Cercidium floridum

Blue palo verde and the more cold-tolerant foothill palo verde (*C. microphyllum*) are two of the best-loved desert trees. Whether covered with yellow flowers, as they are in April, or with delicate lime-green leaves, as in the spring and fall rainy seasons, these multitrunked trees always look elegant because of their smooth, light green trunks. Blue palo verdes have pale, blue-green trunks and fragrant golden flowers. Foothill palo verdes have lime-green trunks and creamy flowers that bloom two weeks later than blue palo verde. The fruits are very short beanpods that ripen in midsummer.

J F M A M J J A S O N D

Propagation is easy from fresh, untreated seed planted in October. Germination should occur in two to three weeks. Blue palo verde grows in the hottest deserts along washes, so it is tolerant of extra watering. It must always be well drained, but it loves to be planted in a fast-draining swale with lots of irrigation heads.

Bees use the nectar from flowers to make honey and the flowers also attract butterflies and the tarantula hawk wasp. Doves, pack rats, and other desert wildlife eat the seeds.

tarantula hawk wasp

FIELD NOTES
- *Zones 9–10*
- *Adapted to southern WC*
- *Drought deciduous and cold deciduous*
- *15–30' (4.5–9 m)*
- ☀ *Full sun*
- ◌ *Moderate*

183

Desert Willow

Chilopsis linearis

Desert willow is a graceful, multitrunked tree.

Though not a true willow, it has slender, willow-like, dark green leaves and very showy flowers. These blooms are tubular with flaring mouths and come in white, pink, rose, and burgundy, as well as two-tone combinations—sometimes with ruffles. Desert willow flowers from spring until fall, after rain, and flowering is almost continuous with irrigation. The fruits are string bean-sized and last through winter.

Germination is easy from fresh seed. If you want a particular color, try dormant cuttings or semihardwood cuttings taken in late summer. There are many pretty selections available in nurseries. Desert willow is fast growing and quickly grows into a tree, but benefits from careful pruning. In desert gardens, place it in a swale and channel rain runoff from your roof or driveway to it.

The flowers are a summer boon to hummingbirds, butterflies, and bees. The leaves are sometimes eaten by the Texan crescentspot butterfly in late spring and early summer. Pigeons and doves eat the flaky seeds.

J F M A M J J A S O N D

FIELD NOTES

■ Zones 7–10

■ Winter deciduous, sometimes drought deciduous

■ 15–40' (4.5–12 m)

☀ Full sun

💧 High

Texan crescentspot butterfly

Golden Ball Lead Tree

Leucaena retusa

Golden ball lead tree is a small, multitrunked tree. After rain, but especially in spring or fall, it is covered with golden globes of sweetly scented flowers. The leaves, divided into lime-green leaflets, cast light shade and appear to dance in even the slightest breeze. By late summer, the long, flat, pale green beanpods have ripened to tan, twisting open to release the seeds.

Gather the fresh seed just before the pods split open. Air-dry for two days, and dust with a natural insecticide, such as Sevin, before storing them in a jar at room temperature. Sow the untreated seed in fall as soon as the weather starts to cool down or in spring after all danger of frost is past. Any well-drained soil is acceptable, but rich, moist soil encourages too-rapid growth and weak branches. This is an excellent courtyard tree.

The fragrant flowers are a source of nectar for butterflies and bees. Insects and songbirds eat the seed.

The broad-billed hummingbird and the varied bunting like to nest in its branches. The foliage and pods are a favorite browse of deer.

J F M A M J J A S O N D

FIELD NOTES
- Zones 8–10
- Winter deciduous
- 15–25' (4.5–7.5 m)
- Full sun to a little shade
- High

bordered patch butterfly

Honey Mesquite

Prosopis glandulosa var. glandulosa

A mesquite tree in winter is a sculpture of dark, twisting trunks and knobby, thorny, black branches. Good pruning is important to make this phase as beautiful as possible. In late spring, delicate, feathery leaves are soon joined by fuzzy fingers of yellow to cream flowers that are fragrant close up. Beanpods follow in late summer. Usually mottled purple and cream, the pods are occasionally a brilliant ornamental red.

J F M A M J J A S O N D

Honey mesquite is native to the Chihuahuan Desert. Torrey mesquite (*P. glandulosa* var. *torreyana*) is its native equivalent in the Sonoran, Mojave, and Lower Colorado deserts. Both have light green leaves. Propagation is by fresh or scarified seed or by root cuttings. Drainage must be good, but extra moisture is appreciated; in a desert garden a swale is ideal.

The flower nectar attracts bees and butterflies, including the great purple hairstreak. Birds and mammals eat the beans, which are sweet-tasting and rich in protein. The roots provide shelter for burrowing animals. Many birds use mesquite for nesting.

FIELD NOTES

- Zones 8–10
- Winter deciduous
- 20–35' (6–10.5 m)
- ☀ Full sun
- ◗ Moderate

great purple hairstreak butterfly

beanpods

186

Joshua Tree

Yucca brevifolia

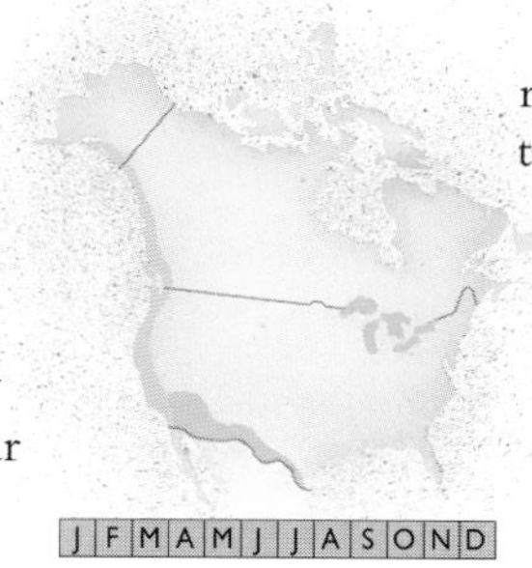

All the shrub yuccas and tree yuccas native to the Desert Southwest are important to wildlife. Joshua tree, the tree yucca of the Mojave Desert, has numerous branches and short leaves. Every few years, clusters of white flowers appear in the late spring followed by hard, tan, seed capsules.

Yuccas can be grown from fresh seed harvested before the capsules split open. Press the seed into sand so that the tops are barely showing, and then lightly dust them with more sand. If you can't plant the seed immediately, store it in moist sand in the

J F M A M J J A S O N D

refrigerator. Yuccas germinate best in temperatures ranging from 60–70°F (16–21°C). Good drainage is essential, and it is very important not to overwater.

Seedlings are eaten by mule deer, rabbits, wood rats, and ground squirrels. The branches of the Joshua tree provide nesting for songbirds, doves, and lizards. Desert night lizards climb the tree looking for insects. Woodpeckers and flickers carve homes in dead trunks. As with the other yuccas, the flowers are pollinated by a small, white yucca moth that deposits eggs in the flower. The larvae eat some of the seeds but leave enough to ensure a continuing supply of yuccas. Giant skipper butterfly larvae feed on the stems and leaves.

FIELD NOTES
- Zone 8
- Adapted to southern WC
- Evergreen
- 20–30' (6–9 m)
- Full sun
- Low

desert night lizard

187

Yellow columbine (left); Mexican redbud (right)

Additional species

Plant name/flower color	Zone	Height	Visitors/attraction/season
CACTI Fishhook barrel cactus (orange) *Ferocactus wislizenii*	8–10	2–11' (60–330 cm)	bees/flowers/summer mammals/fruits/fall
HERBACEOUS FLOWERS Yellow columbine (yellow) *Aquilegia chrysantha*	8–10	1–3' (30–90 cm)	hummingbirds/flowers/spring
Blackfoot daisy (white) *Melampodium leucanthum*	7–10	1' (30 cm)	butterflies/flowers/all year
Bush muhly (pink) *Muhlenbergia porteri*	4–10	1–3' (30–90 cm)	birds/seeds/fall mammals/browse/summer
Mexican plumbago (white, blue) *Plumbago scandens*	9–10	2–3' (60–90 cm)	butterflies/flowers/spring, fall
Skeletonleaf goldeneye (yellow) *Viguiera stenoloba*	8–10	2–4' (60–120 cm)	butterflies/flowers/after rain caterpillars/leaves/spring
Dwarf white zinnia (white) *Zinnia acerosa*	9–10	3–6" (8–15 cm)	butterflies/flowers/spring–fall
SHRUBS Flame acanthus (orange) *Anisacanthus quadrifidus* var. *wrightii*	8–10	2–6' (60–180 cm)	hummingbirds/flowers/summer–fall
Littleleaf cordia (white) *Cordia parviflora*	9–10	6–12' (1.8–3.7 m)	butterflies/flowers/spring birds, mammals/fruits/summer
Mormon tea (yellow) *Ephedra trifurca*	8–10	4–15' (1.2–4.5 m)	birds/cones/fall bighorn sheep/stems/all year
Apache plume (white) *Fallugia paradoxa*	8–10	4–6' (1.2–1.8 m)	butterflies/flowers/spring–fall
Red yucca (coral) *Hesperaloe parviflora*	8–10	2–5' (60–150 cm)	hummingbirds/flowers/spring–fall
Desert lavender (lavender) *Hyptis emoryi*	9–10	8–12' (2.4–3.7 m)	birds, butterflies/flowers/spring birds/seeds/summer
Creosote bush (yellow) *Larrea tridentata*	8–10	6–10' (1.8–3 m)	quail/cover/all year birds, mammals/seeds/fall–winter
Texas sacahuista (white) *Nolina texana*	8–10	2' (60 cm)	bees/flowers/spring mammals/seeds/fall
TREES Netleaf hackberry (n/a) *Celtis reticulata*	8–10	15–75' (4.5–23 m)	caterpillars/leaves/spring birds/seeds/winter
Mexican redbud (pink) *Cercis canadensis* var. *mexicana*	8–10	20' (6 m)	bees/flowers; caterpillars/leaves/spring birds/seeds/fall–winter
Ironwood (pink) *Olneya tesota*	9–10	15–30' (4.5–9 m)	bees/flowers/summer mammals/beans/fall
VINES Thicket creeper (n/a) *Parthenocissus inserta*	4–10	10' (3 m)	birds/fruits/fall mammals/fruits/fall
Canyon grape (n/a) *Vitis arizonica*	8–10	50' (15 m)	birds/mammals/fruits/fall sphinx moth caterpillars/leaves/spring

The Prairies

THE PRAIRIES
The Prairie Garden

The Prairie region forms the heart of North America. The Prairies were once a vast grassland, with taller grasses to the east, and progressively shorter grasses to the west. In spite of the low annual rainfall, the rich soil proved conducive to farming on a scale never before seen. The resulting devastation of the prairie grasslands in the interests of agriculture was no less significant and thorough than the current destruction of the great rainforests of the world. Today, only tiny remnants of true prairie grassland remain, mostly in preserves and parklands. Fortunately, renewed interest in the plants of the prairie now influences the development of gardens in this region.

The prairie garden recalls the wide open aspect of the plains. Even in the small garden, the effect of distance can be achieved by leaving the margins of the garden open to the far-off views. Reflecting the importance of water in the prairies, the gardener may place a small pond near the patio, within view of the house. One or two riparian trees shade the house, and provide excellent habitat for a broad range of birds who prefer the shelter of trees to the open plains. The bur oak (*Quercus macrocarpa*) or the common hackberry (*Celtis occidentalis*) offer food as well for squirrels and cedar waxwings. The downy hawthorn (*Crataegus mollis*) and chokecherry (*Prunus virginiana*) provide a dense screen for privacy and to hide the nearby highway. Their flowers attract butterflies in late spring, their fruits bring birds, particularly during the fall migration, their fall foliage brightens the landscape, and their dense growth helps buffer cold, winter winds.

A small lawn surrounding the patio serves the needs of a children's playground, as well as a foreground to the simulated prairie beyond. Here, a mass of native grasses, mostly sideoats grama (*Bouteloua curtipendula*) and little bluestem (*Schizachyrium scoparium*), provide a fabric of delightful texture, color, and movement, with drifts of native perennials offering additional color and textural variety. Monarchs and other butterflies are attracted to purple coneflower (*Echinacea purpurea*) and Maximilian's sunflower (*Helianthus maximiliani*), interspersed with butterfly weed (*Asclepias tuberosa*) on drier ground, or wild bergamot (*Monarda fistulosa*) where there is more moisture. In fall, American goldfinches flock to feed on the golden seed heads of the grasses, ripening alongside the late flowers of goldenrods (*Solidago* spp.).

Big Bluestem

Andropogon gerardii

Big bluestem is a tall, long-lived bunchgrass of the eastern Great Plains and wet areas farther west. It is one of the dominant grasses of the tallgrass prairies, once found in almost pure stands over vast areas. In midsummer, big bluestem rapidly develops a tall, reddish stem, topped with three-pronged flower heads. These flower heads look surprisingly like a turkey's foot, and from this comes another common name, turkeyfoot bluestem. Native American names usually referred to the beautiful, burgundy-red fall color of this grass.

Big bluestem is widely adapted to a range of soils. Propagate it from seed, which won't germinate until the soil is warm in spring. You can plant it at other times if the effects of wind and water erosion can be controlled.

Songbirds and hoofed browsers, including white-tailed deer, are the main wildlife users of this grass. Sparrows and juncos will perch on the stalks to pluck and eat the seeds. Big bluestem is an important host of butterflies, including the regal fritillary. Adult regal fritillaries feed on the nectar of milkweeds and thistles within the tallgrass ecosystem.

J F M A M J J A S O N D

grass head

white-tailed deer

FIELD NOTES
- Zones 3–8
- Perennial
- 3–9' (1–3 m)
- ☀ Full sun
- ◐ Moderate–high

192

Blue Grama

Bouteloua gracilis

Extremely drought-tolerant, blue grama is a long-lived, warm-season bunchgrass. The seed heads are very distinctive, resembling eyebrows on stalks. The blue–green color inspires the name blue grama, a common color of plants of hot, sunny, dry conditions. This lighter color reflects some of the sun's heat, allowing photosynthesis to occur for longer periods.

Blue grama is widely adapted to a variety of soils. Propagate it from seed, which requires warm soil to germinate—wait until catalpa trees are about to bloom. It is best not to plant blue grama after mid-July at latitude 40°N to allow the plants time to become established before winter.

You can use blue grama as a wonderful, informal lawn. You can either mow it, or leave it unmowed and plant a large number of short-grass prairie wildflowers such as chocolate flower (*Berlandiera lyrata*), gayfeather (*Liatris punctata*), and poppy mallow (*Callirhoe involucrata*).

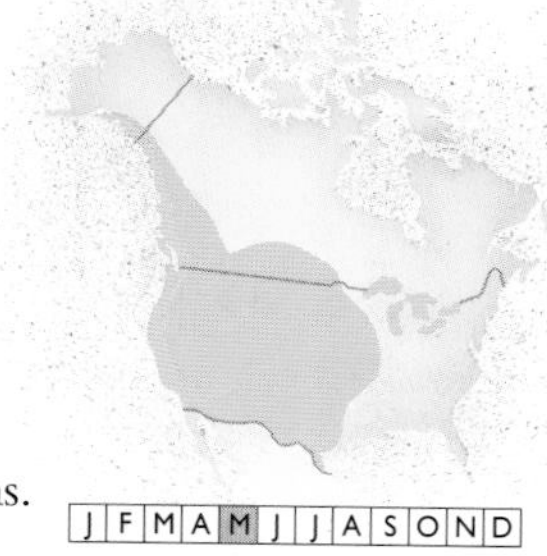

Wild turkeys, brown-capped rosy finches, and chestnut-collared longspurs are reported to eat blue grama seeds. Mice and voles gather the seeds for their winter storehouse.

| J | F | M | A | M | J | J | A | S | O | N | D |

FIELD NOTES
- Zones 3–9
- Seed heads, generally, 6–12" (15–30 cm); leaves, generally 3–6" (7.5–15 cm)
- ☀ Full sun essential
- ◯ Low

chestnut-collared longspur

193

Little Bluestem

Schizachyrium scoparium

Little bluestem is a warm-season, long-lived bunchgrass. The fall color is an attractive reddish brown, and the plant is topped with decorative, fuzzy seed heads. It is especially beautiful when backlit by the sun. Plant it as a single specimen or as a meadow planting.

J F M A M J J A S O N D

To survive the severe dry conditions so common in midgrass prairies, little bluestem develops extensive roots that reach deep below the surface.

Little bluestem is widely adapted to a range of soils. Propagate it from seed, which won't germinate until the soil is warm. You can plant it at other times of the year if the effects of wind and water erosion can be controlled.

Rosy finches and juncos, as well as chipping, field, and tree sparrows, eat the seeds of little bluestem. Meadowlarks nest in the midgrass prairies, where little bluestem grows, and dusky skipper butterfly caterpillars overwinter in tube tents above the base of little bluestem clumps.

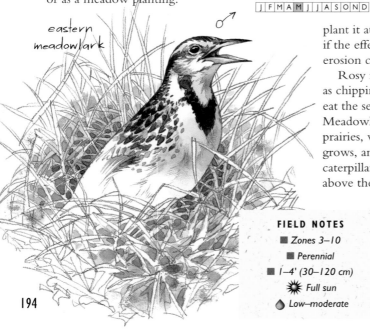

eastern meadowlark ♂

grass head

FIELD NOTES
- Zones 3–10
- Perennial
- 1–4' (30–120 cm)
- Full sun
- Low–moderate

Pearly Everlasting

Anaphalis margaritacea

Pearly everlasting produces masses of small, white flowers beginning in midsummer. The seed heads that form later extend the showy period considerably. The silvery white leaves are attractive throughout the season. The species name *margaritacea* is from the Greek for pearl-like. The flowers dry well for bouquets—thus, the name everlasting. Cut the flowers when the centers begin to show, and then put them in water briefly before hanging them upside down for drying.

J F M A M J J A S O N D

Pearly everlasting is adapted to a wide range of moisture conditions, from quite dry to quite damp, so long as the drainage is good. It is an unusual example of a plant with silvery leaves that tolerates moist conditions. Propagate it from seed or division.

Butterflies, bees, and hover flies relish the nectar of pearly everlastings. American painted lady butterflies lay their yellowish-green eggs on pearly everlastings, and later, the colorful caterpillars form nests of silk and leaves on these same plants. American painted ladies may overwinter as adults.

hover fly

> **FIELD NOTES**
> ■ *Zones 3–9*
> ■ *Perennial*
> ■ *To 4' (1.2 m)*
> ✺ *Best in full sun, but adapted to some shade*
> ◌ *Low–moderate*

195

Butterfly Weed

Asclepias tuberosa

Butterfly weed is widely adapted to a range of soils, from very dry to very moist. You can try propagating it from seed, but results are often poor; otherwise, propagate it from root division. You can transplant nursery-grown plants at any time, but unfortunately, the plants' success is often uncertain. The flowers do well in bouquets. Unpicked flowers will develop showy seedpods with silky seeds. Be patient when weeding in spring, because butterfly weed is slow to emerge.

Butterfly weed is well named, for it is very attractive to many butterflies, as well as bees. The broad, flat flower heads provide generous landing platforms for butterflies, with lots of individual flowers offering tiny "pots" of sweet nectar. After their caterpillars feed on various milkweed plants, monarch and other milkweed butterflies, including the queen, develop a bitter taste that helps protect the adults from predators. The chrysalises are beautiful, jade-green, and studded with gold dots.

| J | F | M | A | M | J | J | A | S | O | N | D |

milkweed bug

FIELD NOTES

■ Zones 4–9

■ Perennial

■ 2–3' (60-90 cm)

☀ Full sun

◗ Low–moderate

queen butterfly caterpillar

Purple Coneflower

Echinacea purpurea

Purple coneflower is an easy-to-grow, showy, long-lived flower that is native in the tallgrass prairie region.

It is adapted to a wide range of soils. Propagate it from seed: sow seed in spring or fall; no special treatment is required. The plant was named after the Greek word *echinos*, for hedgehog —a reference to the prickly, showy, flower centers. Purple coneflower makes attractive bouquets, either with fresh flowers or by using the dried, prickly seed heads in dry arrangements.

J F M A M J J A S O N D

The flowering of this plant almost epitomizes midsummer, and it seems as if all the butterflies of midsummer, such as the red admiral, like it. Fritillaries are especially frequent visitors. Birds such as the American goldfinch will feed on the seeds in late summer and fall. A big plus is that deer do not eat this plant.

The plant contains a compound toxic to mosquitoes and houseflies, and efforts are under way to use another insect-inhibiting compound as a natural insecticide on sunflower crops.

red admiral butterfly

FIELD NOTES
- ■ Zones 3–9
- ■ Perennial
- ■ 2–3' (60–90 cm)
- ☀ Full sun
- 🌢 Moderate

197

Maximilian's Sunflower

Helianthus maximiliani

Maximilian's sunflower makes a real show in the fall landscape. It blooms late, often into October, and the numerous stems lined with many bright yellow flowers create quite a display. It is named after Maximilian Alexander Philipp of Wied-Neuwied, who collected many plants on an expedition up the Missouri River in 1833 and 1834. This is a long-lived perennial that is considered native throughout the Great Plains, but it is most associated with the tallgrass prairies, where there is relatively ample moisture. In dry locations, this sunflower is much smaller than on moist sites. It has become a significant part of northern New Mexico landscaping, where it adds to the dramatic, yellow fall flowers of rabbitbrush (*Chrysothamnus* spp.), various senecios, and the wonderful blue of Bigelow's asters (*Machaeranthera bigelovii*).

Maximilian's sunflower is adapted to a wide range of soils. Propagation is usually by seed.

The flowers attract numerous butterflies. Colorful lazuli buntings and white-crowned sparrows feed on the seeds, which have been grown as a source of excellent birdseed.

J F M A M J J A S O N D

FIELD NOTES

■ Zones 2–8

■ Perennial

■ 5–10' (1.5–3 m)

☀ Full sun

◐ Moderate

♂

white-crowned sparrow

198

Bush Morning-glory

Ipomoea leptophylla

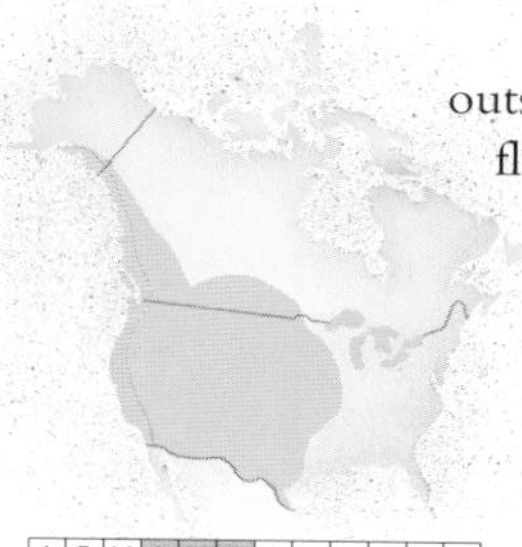

Bush morning-glory is a shrub-like perennial with fine, narrow leaves and extremely showy, pink, funnel-shaped flowers from late May to August.

Bush morning-glory prefers sandy, well-drained soil. Propagate it from seed or from root-crown division. It prefers dry conditions—no more than 12–14" (300–350 mm) average annual rainfall. Bush morning-glory will grow in most zones, provided the preferred soil and dry climate are present.

The deep, tubular flowers of this plant are pollinated mostly by bees, but ants seem fond of the plant, too. The ants are likely acting as guards, warding off some other insect that would "waste" valuable pollen or nectar. Ants themselves are considered pollination scoundrels, because their slippery bodies don't carry pollen well. They like nectar so much, however, that many plants have developed external nectaries to attract them to areas outside the pollen-bearing parts of the flowers, where they can fend off other would-be robbers. Butterflies, including the cloudless sulfur, are also attracted to the flowers of bush morning-glory.

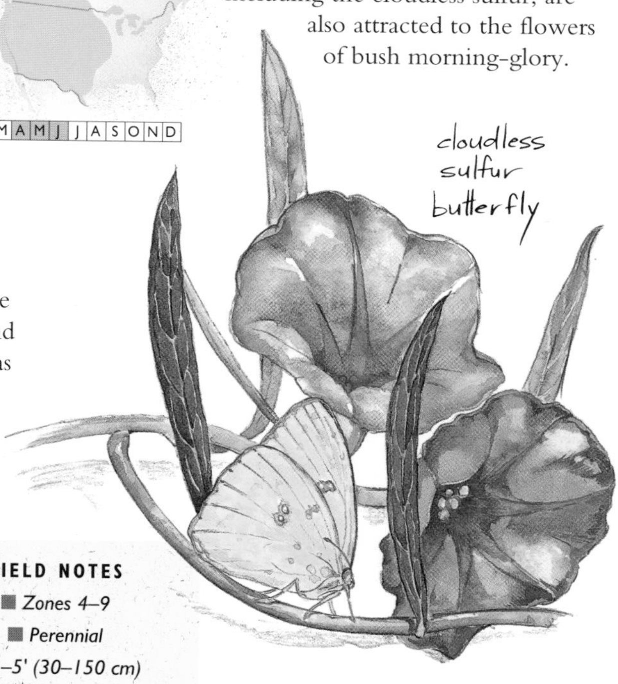

J F M A M J J A S O N D

cloudless sulfur butterfly

FIELD NOTES

■ Zones 4–9

■ Perennial

■ 1–5' (30–150 cm)

☼ Full sun

◊ Low

199

Dotted Gayfeather

Liatris punctata

This is a very ornamental flowering plant, useful in dry gardens or in shortgrass meadows. It has a moisture-storing rootstock from which it develops erect stems covered with pretty, pink flowers from late summer well into fall.

mason bee

Individual plants probably bloom for no more than two weeks, but other plants keep the show going for nearly three months. Gayfeathers are unusual in that flowering begins at the top of the flower spike, then progresses downward (most plants with flowers in vertical spikes bloom from bottom to top). Dotted gayfeather makes an attractive cut flower for bouquets.

Gayfeather prefers well-drained soil and will grow in most zones provided it has dry conditions—more than 12–14" (300–350 mm) average annual rainfall may cause problems.

Propagate it from seed or root division; sow seed in very early spring or in late fall (stratify seed for spring planting); plant root divisions in the same periods.

When dotted gayfeather is in bloom, its nectar attracts bees and numerous butterflies, including the regal fritillary.

J F M A M J J A S O N D

regal fritillary butterfly

FIELD NOTES
- Zones 3–10
- 6–24" (15–60 cm)
- Full sun
- Low

200

Wild Bergamot

Monarda fistulosa

Wild bergamot is an upright plant with fragrant foliage on square stems, topped by clusters of pretty, tubular summer flowers. It is common in the Rocky Mountain foothills and in the relatively wet, eastern tallgrass prairies. Wild bergamot, wild monarda, bee balm, horsemint, Oswego tea, purple bergamot, oregano, plains bee balm, and fern mint—this plant has many names! The most interesting, however, may be the Pawnee name *tsakus tawirat* (shot-many-times-and-still-fighting), which no doubt refers to the medicinal properties of this plant, such as its antibacterial and antifungal properties.

Like so many of the mints, wild bergamot is likely to spread rapidly with fertile soil and plenty of water. It is adapted to a wide range of soils but does not like dry conditions.

J F M A M J J A S O N D

Propagate it from seed, or from plant division in early spring, or in fall after flowering.

A range of butterflies, hummingbirds, bees, and diurnal moths, such as the hummingbird moth, will be attracted to wild bergamot flowers.

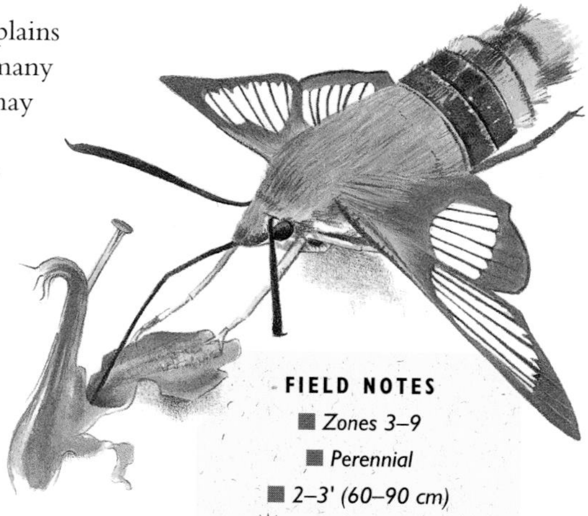

hummingbird moth

FIELD NOTES
- Zones 3–9
- Perennial
- 2–3' (60–90 cm)
- Full sun–part shade
- Moderate–high

201

Ozark Sundrop

Oenothera missouriensis

Ozark sundrop is a long-lived perennial with large, fragrant, lemon-yellow flowers that tend to open near sunset and close before noon the following day. The blooming season is from spring to fall. Although the genus *Oenothera* is often referred to as evening primrose, in reality, some species open in the evening; and others, in the morning. Oil of evening primrose is said to be the richest natural source of unsaturated fatty acids and is widely mentioned in natural-food publications. Scientific research has also shown it to be useful in cholesterol reduction.

J F M A M J J A S O N D

Ozark sundrop is adapted to a wide range of soils. Propagation is usually from seed, although you can propagate it from root division.

All species of *Oenothera* are pollinated primarily by various night-flying moths, or, for those flowering during the day, by hummingbirds and butterflies, all attracted by the nectar. Tree crickets are attracted to caterpillars that feed on the plants.

FIELD NOTES
- Zones 4–8
- Perennial
- 6–12" (15–30 cm)
- ☀ Full sun
- ◌ Moderate–high

American tree cricket

202

Foxglove Penstemon

Penstemon cobaea

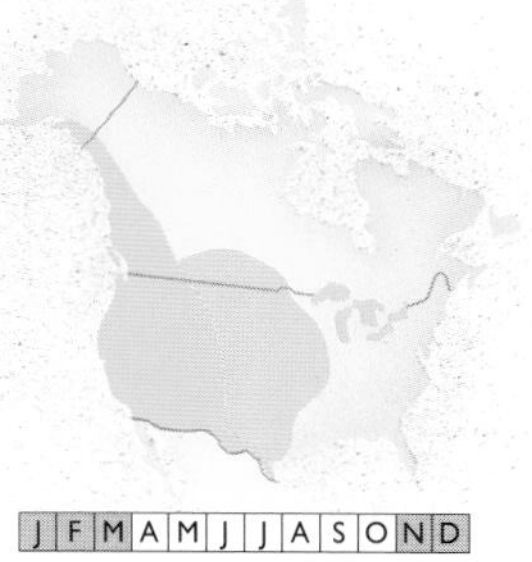

Penstemons make up the largest genus of wildflowers in North America, and they are native only to North America. There are tall ones, short ones, and some for dry as well as wet locations. The spring flowers can be white, red, blue, or yellow.

The genus name *Penstemon* refers to the five stamens they possess, but the fifth stamen has no anther. Instead, it serves as a barrier to crawling insects that might rob nectar without distributing pollen to flowers on distant plants. Foxglove penstemon is among the showiest penstemons, and is one of the longest-lived in gardens. The flowers are white and very large, with pretty, reddish lines.

J F M A M J J A S O N D

bumblebee on
P. cardinalis

Plant foxglove penstemon in well-drained soil. Propagate it from seed (which will probably need stratification). Sow seed in fall, winter, or early spring.

Foxglove penstemon is pollinated mostly by bumblebees, which "muscle" their way into the large, tubular flowers. Hummingbirds will also be attracted to the flowers.

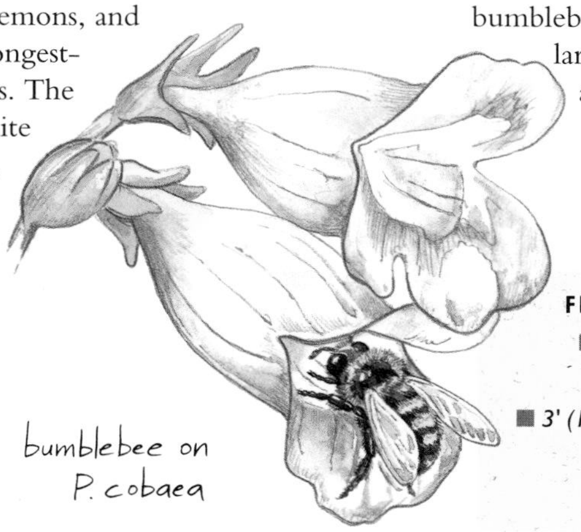

bumblebee on
P. cobaea

FIELD NOTES
- Zones 4–9
- Perennial
- 3' (1 m) when in bloom.
- ☀ Full sun
- ◊ Moderate

203

Blue Sage

Salvia azurea

Blue sage is a tall, long-lived plant with appealing, light blue flowers in spike-like clusters. Flowers typically bloom from late August to early October. It is generally associated with tallgrass prairie plant communities, where it is particularly showy amidst the various grasses and yellow-flowered Maximilian's sunflower (*Helianthus maximiliani*).

In gardens, blue sage usually stands upright better when grown in dry conditions, but it is adapted to a wide range of soils. Propagate it from seed; you can also propagate it from root divisions and cuttings. It will self-sow and naturalize in garden areas. Pinching back the stems several times before early July results in more stems and, therefore, more flowers. *S. azurea* var. *grandiflora* is a natural variety with larger, lighter blue flowers.

A variety of insects, including bees, hover flies, and butterflies, will visit blue sage. Adult monarch butterflies have shown considerable interest in the flowers, which bloom from late summer to early fall. Hummingbirds in the area will also discover the nectar in the blue flowers.

| J | F | M | A | M | J | J | A | S | O | N | D |

hover fly

eastern box turtle

FIELD NOTES
- Zones 4–10
- Perennial
- 3–4' (90–120 cm)
- Full sun
- Low–moderate

204

Goldenrods

Solidago spp.

There are many species of goldenrod in Europe, Asia, and North America. They are widely adapted to a range of soils and grow in hot, cold, dry, and wet areas. They have a generally similar appearance and can be quite showy, with numerous stems topped by clusters of yellow flowers.

Despite their showiness, some are very invasive, while others remain compact. European gardeners have developed some of the best new garden cultivars from the North American S. canadensis. 'Baby Gold' is one of these.

Propagate goldenrods from seed, division, or stem cuttings. Sow seed or plant young specimens in spring. Some seed may need stratification.

Goldenrods tend to bloom from midsummer into fall, and are good butterfly-attracting plants. While the tiny tubes of goldenrods' flowers contain only minuscule amounts of nectar, the large number of flowers in each cluster compensates for this. Monarch, orange sulfur, and red admiral butterflies will all be attracted to goldenrod flowers.

| J | F | M | A | M | J | J | A | S | O | N | D |

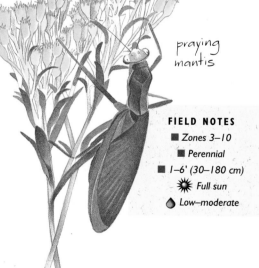

praying mantis

FIELD NOTES
- Zones 3–10
- Perennial
- 1–6' (30–180 cm)
- Full sun
- Low–moderate

205

Verbena canadensis 'Homestead Purple'

Rose Verbena

Verbena canadensis

Rose verbena is a low-growing, short-lived perennial generally associated with the eastern tallgrass prairies. The variable bloom time ranges from spring to fall, with flowers that are usually more pink than blue. Various verbenas have been used for many medicinal purposes, including stomach troubles, respiratory inflammations, fevers, and, perhaps best of all, "a feeling of relaxed well-being".

In the garden, rose verbena will self-sow and naturalize if the ground is open and there is little competition from other plants. It prefers a well-drained soil. Propagate it from seed or from cuttings.

J F M A M J J A S O N D

This verbena is a typical butterfly-pollinated flower. Its flat-topped flowers make landing easy for butterflies, and butterflies have the long tongues needed to reach deep into the flowers for nectar. Bees generally do not have mouthparts long enough for verbena flowers, but some bees have special forelegs that allow them to collect pollen by inserting them into the tubes. Verbenas support all phases (egg, caterpillar, chrysalis, and adult) of the attractive Theona checkerspot butterfly.

FIELD NOTES
- Zones 5–10
- Perennial
- 6" (15 cm)
- Full sun
- Low–moderate

great spangled fritillary butterfly

Chokecherry

Prunus virginiana

Chokecherries are tall, thicket-forming shrubs with attractive, spring flower clusters and very useful, midsummer fruits. Native Americans used the bark to treat coughs, fevers, and arthritis. In mainstream modern medicine, chokecherry is used mainly in cough medicine for its mild, sedative properties and pleasant taste. The name chokecherry is well deserved: the berries are very astringent when eaten raw. They are, however, delicious as a syrup topping.

Chokecherry tolerates a range of soils but prefers well-drained soil. Propagate it from seed sown in fall, or stratify seeds till spring.

Tiger swallowtail, two-tailed swallowtail, and Lorquin's admiral butterflies use chokecherries for their chrysalis phases. The chrysalises of the swallowtails are usually green or brown and mimic pieces of leaf or wood.

A favorite fruit of raccoons, chokecherries are also popular with fruit-eating birds, such as the eastern kingbird and, particularly, the robin. Even a single plant can become the center of a very conspicuous feeding frenzy.

J F M A M J J A S O N D

eastern kingbird ♂

tiger swallowtail cocoon

FIELD NOTES
- Zones: 2–8
- Deciduous
- 10–25' (3–8 m)
- ☀ Full sun
- ◊ Moderate–high

207

Smooth Sumac

Rhus glabra

Smooth sumac is a medium-sized, thicket-forming shrub. Despite a dense and leafy appearance in summer, when smooth sumac drops its compound leaves in fall, the result is a shrub with few branches, which creates a distinctive branching pattern in the winter landscape, but limits its use as a year-round screen. Before dropping, the large, divided leaves turn brilliant shades of red, orange, and purple, lighting the fall landscape.

J F M A M J J A S O N D

Smooth sumac is widely adapted to a range of soils. Propagate it from scarified seed, cuttings, or root pieces collected in midwinter. Scarify the seeds in sulfuric acid for 50–80 minutes.

Thirty-two species of bird eat the reddish sumac berries, which last from fall into winter, including wild turkeys, pheasants, robins, Steller's jays, flickers, quails, and downy woodpeckers. The thicket-forming habit creates very good cover for both birds and mammals, including chipmunks and rabbits. Rufous-sided towhees seem to be especially fond of foraging beneath these sumacs.

FIELD NOTES
- Zones 2–9
- Deciduous
- 9–15' (2.7–4.5 m)
- Full sun or light shade
- Low–moderate

downy woodpecker ♂

Wood's Rose

Rosa woodsii

Wood's rose is a low-growing shrub that forms patches in grassland areas. The rosy red to orange fall color can be quite attractive. The pink flowers are pretty, and the fall fruits (hips) are both attractive and useful, being high in vitamin C.

Wood's rose is widely adapted to a range of soils. Propagate it from fresh seed, which may not germinate until the second year after planting. You can also use root divisions and cuttings of green wood.

Deer are especially fond of rose buds and twigs, so be aware that after browsing on them, they may move on to other plants in the garden.

Rose hips remain on the plants for a long time, sometimes until spring, and are an important source of food when other sources are gone. Birds are attracted to the hips. Pollination of Wood's rose is usually by native bumblebees rather than European honeybees. So many interesting insects are attracted to wild roses for the nectar, pollen, and foliage that full appreciation of these roses might require an insect field guide!

| J | F | M | A | M | J | J | A | S | O | N | D |

black-and-yellow argiope

fall fruit (hip)

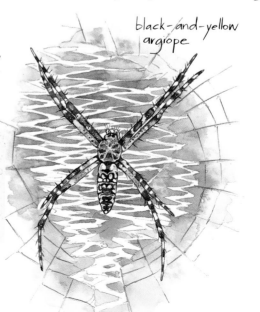

FIELD NOTES
- Zones 3–8
- Deciduous
- Usually less than 3' (1 m)
- Full sun or light shade
- Moderate

209

Soapweed

Yucca glauca

Soapweed, or soapwell, is a tough, evergreen shrub with sharp, bayonet-like leaves. The showy, greenish white flowers are produced in large clusters along sturdy stems in May or early June. Found in the western half of the Great Plains, soapweed has an extensive root system that allows it to survive on particularly dry sites. Native Americans pounded the roots to make a soap, giving rise to the common names.

J F M A M J J A S O N D

Soapweed is adapted to most dry soils. Propagate it from seed, or by removing and rooting the offsets that develop close to the ground.

Yuccas all over the continent have developed a close association with the pronuba, or yucca, moth. The female moths collect pollen from one flower and pack it onto the stigma of another flower, usually on another plant. This ensures cross-pollination and the production of a good crop of seeds. She also lays eggs inside the flower's ovary; the moth's larvae feed on the seeds developing inside the ovary, but leave enough uneaten to guarantee another generation of soapweeds.

Strecker's giant skipper butterflies glue their green eggs to the undersides of soapweed leaves; later, the caterpillars feed on the roots.

yucca moth

FIELD NOTES
- Zones 4–8
- Evergreen
- 3'+ (1 m+)
- ☀ Full sun
- 💧 Low

210

Common Hackberry

Celtis occidentalis

With adequate water, common hackberry is an attractive, large tree. In the western high plains, where the climate is relatively dry, it often grows in thickets as a shrub.

Common hackberry will tolerate a range of soils. Propagate it from stratified seed, as cuttings do poorly. Water regularly to establish the plant after transplanting.

The seeds of the common hackberry attract cedar waxwings, which sometimes perform a charming ritual of passing one of the hard, round seeds back and forth until one of the birds finally swallows it. Both a parent feeding young and a pair performing courtship rituals will exchange seed in this way. Hackberry trees host numerous insects that attract many birds, including cedar waxwings. Mourning cloak butterflies feed on hackberry sap. Hackberry trees are the sole host of hackberry butterfly caterpillars, which resemble pretty, green slugs with ornamental markings. Raccoons, bears, coyotes, skunks, foxes, and squirrels are known to eat hackberry seeds.

J F M A M J J A S O N D

hackberry butterfly caterpillar

FIELD NOTES

- Zones 2–9
- Deciduous
- 40–60' (12–18 m) (tree); 10–30' (3–9 m) (shrub)
- ☀ Full sun
- ◊ Moderate

hackberry butterfly cocoon

Downy Hawthorn

Crataegus mollis

Hawthorns have a long history in legends. Most famous is that Christ's crown of thorns was made of hawthorn. Downy hawthorn is an attractive, small tree with pretty bark, spring flowers, attractive, red fall fruits, and beautiful red or yellow fall color. The name hawthorn refers to the plant's thorns and haws (fruits).

Downy hawthorn is adapted to a range of soils. Propagation is difficult, so buy nursery-grown specimens to plant.

The thorniness of hawthorns, and the tendency to grow in thickets, create excellent cover for wildlife. Rufous-sided towhees can be heard more often than seen, as they scratch around on the ground foraging for food in hawthorn thickets. Fox sparrows and cedar waxwings are among the surprisingly few wildlife species that eat hawthorn fruits, which is one reason the fruits are ornamental for so long. Striped hairstreak, northern hairstreak, white admiral, and red-spotted purple butterflies use hawthorns as hosts for their egg, caterpillar, and chrysalis phases.

J F M A M J J A S O N D

white admiral butterfly cocoon

FIELD NOTES
- Zones 3–8
- Deciduous
- 20–30' (6–9 m)
- Full sun
- Moderate

♂

eastern rufous-sided towhee

Eastern Cottonwood

Populus deltoides

Found throughout the eastern two-thirds of the continent, the eastern cottonwood is a huge tree that visually marks stream corridors, especially on the western high plains. The brilliant yellow fall color of this easy-to-grow tree heralds the end of the fall color season.

J F M A M J J A S O N D

northern orioles in nest

♀

Eastern cottonwood will tolerate a range of soils. Propagate it from seed, without stratification, or from twigs before spring leaves appear.

The buds, catkins (flowers), seeds, twigs, foliage, bark, and the numerous insects that inhabit the tree, are very important to a wide array of wildlife for shelter and food. Both orchard and northern orioles are especially closely associated with cottonwoods. These colorful, noisy birds often build their intriguing, hanging nests in the topmost branches. Beautiful black-headed grosbeaks frequently project their nonstop, robin-like song from the highest branches. Mourning cloaks, Compton tortoise-shells, white admirals, and red-spotted purples are among the butterflies that use cottonwoods in various phases of their development.

FIELD NOTES

■ *Zones 3–9*

■ *Deciduous*

■ *75–100' (23–30 m)*

☀ *Full sun*

💧 *High*

213

Bur Oak

Quercus macrocarpa

Bur, or mossy cup, oaks are massive, spreading trees with conspicuous, large, dark branches and rounded leaf lobes. The acorns have a distinctive, large cap with "burs", hence, the names bur oak and mossy cup oak. This is a relatively slow-growing tree (compared with cottonwoods, for example), but a long-lived tree—to 300 years or more. It develops just two leaves from stored reserves in the acorn, then grows a taproot up to 3' (1 m) long, before producing any more leaves. The thick bark is fire-resistant.

acorn

J F M A M J J A S O N D

Bur oak is adapted to a range of soils and grows well in clay. Propagate it from seed; stratification for 30–60 days at 41°F (5°C) is suggested.

Bur oak acorns are an important food source for many wildlife species—particularly in the Prairies, where there are few trees—including wild turkeys, eastern fox squirrels, and chipmunks. In all, possibly 500 wildlife species use acorns in North America, including most of the larger seed-eating birds. The greatest value of acorns is in winter, when other food sources are scarce.

eastern fox squirrel

FIELD NOTES
- Zones 3–8
- Deciduous
- 70–80' (21.5–24 m)
- ☀ Full sun
- ◗ Moderate

Trumpet Creeper

Campsis radicans

Trumpet creeper is an extensive, rambling vine with clusters of spectacular, red to orange, trumpet-shaped flowers. The lush, compound leaves appear rather late compared to those of other deciduous plants. The flowers occur in midsummer, in most areas, and continue into fall. Trumpet creeper can cover huge areas—it is not a vine for small spaces. In the wild, it climbs by tangling itself among branches as it rambles through trees and larger shrubs. It also has aerial, root-like holdfasts that help attach the vine to flat surfaces by secreting adhesive substances. In gardens, it climbs best on some sort of natural or artificial support, such as a trellis, fence, or arbor.

Trumpet creeper is widely adapted to a range of soils. Propagate it from seed (stratified for two months at 41°F [5°C] or sown in fall) or from softwood cuttings.

J F M A M J J A S O N D

The flower nectar is very attractive to hummingbirds, such as the ruby-throated hummingbird, which are lured into gardens where they otherwise would rarely be seen. The tangled vines provide shelter for many species of bird.

♀
ruby-throated
hummingbird

FIELD NOTES

- ◼ Zones 5–9
- ◼ Deciduous
- ◼ 30–40' (9–12 m)
- ☀ Full sun
- ◊ Moderate

215

Showy milkweed (left); smooth penstemon (right)

Additional species

Plant name/flower color	Zone	Height	Visitors/attraction/season
CACTI Pincushion cactus (pink) *Coryphantha vivipara*	4–9	4" (10 cm)	bees/flowers/spring
HERBACEOUS FLOWERS Showy milkweed (pink, purple) *Asclepias speciosa*	3–10	3–5' (90–150 cm)	butterflies, bees/flowers/summer monarch caterpillar/leaves/end summer
Day flower (blue) *Commelina dianthifolia*	7–9	6–12" (15–30 cm)	bees/flowers/summer, fall
Sacred datura (white) *Datura meteloides*	6–10	1–3' (30–90 cm)	sphinx moths/flowers/spring–fall
Cutleaf iron plant (yellow) *Haplopappus spinulosus*	3–8	1–2' (30–60 cm)	bees/flowers/fall
Great blue lobelia (blue) *Lobelia siphilitica*	4–9	1–3' (30–90 cm)	bees/flowers/summer, fall
Showy evening primrose (white) *Oenothera speciosa*	5–10	15–24" (38–60 cm)	moths/flowers/spring–fall
Buckley's penstemon (pale pink) *Penstemon buckleyi*	5–8	1½–3' (45–90 cm)	bees/flowers/spring
Smooth penstemon (white) *Penstemon digitalis*	4–9	2–3' (60–90 cm)	bees/flowers/spring
Sidebells penstemon (pink) *Penstemon secundiflorus*	4–7	1–2' (30–60 cm)	bees/flowers/spring
Broom groundsel (yellow) *Senecio spartioides*	4–9	3' (90 cm)	butterflies, bees/flowers/fall
Prairie zinnia (yellow) *Zinnia grandiflora*	5–9	12–18" (30–45 cm)	bees/flowers/spring–fall
SHRUBS Chocolate flower (yellow) *Berlandiera lyrata*	6–9	12–20" (30–50 cm)	bees, butterflies/flowers/summer
Coyote willow (n/a) *Salix exigua*	3–9	3–25' (1–7.5 m)	birds/insects/all year
Silver buffaloberry (n/a) *Shepherdia argentea* (exotic)	2–6	15' (4.5 m)	birds, bears, ground squirrels/ berries/fall, winter
TREES Green ash (n/a) *Fraxinus pennsylvanica*	3–9	66' (20 m)	squirrels/seeds/fall
Hop tree (green) *Ptelea trifoliata*	4–8	25' (7.5 m)	carrion flies/flowers/spring birds/cover/spring–fall
Peach-leaf willow (n/a) *Salix amydaloides*	5–8	50–65' (15–20 m)	birds/insects/all year
Soapberry tree (white) *Sapindus drummondii*	8–10	50' (15 m)	birds/cover/spring–fall
VINES Western virgin's bower (light yellow) *Clematis ligusticifolia*	3–10	3–50' (1–15 m)	bees/flowers/summer

The Northeast

THE NORTHEAST
The Backyard Meadow Garden

The Northeast region stretches north to the Canadian maritime provinces, west to the Great Lakes, and south to northern Tennessee and Virginia. Eastern North America is essentially a vast plain, bisected down the middle by the ancient Appalachian mountain range. To the west of the mountains is a more or less level plateau that extends across to the Mississippi River. East, the land falls away gradually through the hilly piedmont ("foot of the mountain") to the coastal plain—a level stretch of land, only a few feet above sea level, that runs from northern New Jersey to tidewater Virginia. The maritime provinces and New England have no coastal plain; there, the hills descend directly to the rocky coastline.

The backyard meadow garden of the Northeast has trees and shrubs around the perimeter to provide nesting sites and food for wildlife. A large tree in the background may be the majestic white oak (*Quercus alba*), a favorite of migrating warblers, as well as wood ducks, blue jays, deer, and squirrels. Canadian hemlock (*Tsuga canadensis*) is the evergreen accent that provides winter cover for many birds and animals, as well as nesting sites for veeries and juncos.

In front of these trees may be some graceful paper birch trees (*Betula papyrifera*) and sassafras (*Sassafras albidum*). The shrub spicebush (*Lindera benzoin*) is planted for its wash of yellow flowers in spring. All these plants host butterfly larvae: the birch attracts the mourning cloak butterfly, and the spicebush and sassafras attract the spicebush swallowtail butterfly.

Allegheny serviceberry (*Amelanchier laevis*) grows near streams. Its early summer fruits feed red foxes, bluebirds, and orioles. The versatile high-bush blueberry (*Vaccinium corymbosum*) has pink, bell-like flowers in spring and edible, blue berries in summer, which are a favorite of scarlet tanagers. American hornbeam (*Carpinus caroliniana*) attracts myrtle warblers. Its twisted, black trunk will combine with red-osier dogwood (*Cornus sericea*) and winterberry (*Ilex verticillata*) to provide winter interest in the garden.

Meadowlarks nest in grassy meadows among sweeps of black-eyed Susans (*Rudbeckia hirta*) and New England asters (*Aster novae-angliae*). Both these flowers attract monarchs, painted ladies, swallowtails, and other butterflies. So does bee balm (*Monarda didyma*), which grows in the moist soil where the meadow meets the stream.

Wild Columbine

Aquilegia canadensis

In spring, the sprightly red and yellow flowers of wild columbine nod in the breeze on wiry stems. The flower is distinctive for its delicate, spurred or tubular petals that alternate with spreading, colored sepals.

In nature, wild columbine is found nestled in tight crevices on rocky, wooded slopes in dappled shade. It is very adaptable in the garden, however, combining equally well with ferns or other woodland wildflowers in the shade, or with early field flowers in a meadow. In hot climates, it needs protection from the afternoon sun.

Wild columbine propagates readily from seed, which should be planted in spring, although it will take two years for the young plants to reach flowering size. It prefers a well-drained, dry, slightly acid soil.

Long-tongued moths, hummingbirds, and checkerspot butterflies are attracted to the nectar contained in wild columbine's tubular petals, while the columbine duskywing butterfly lays its eggs on the foliage.

J F M A M J J A S O N D

FIELD NOTES
- Zones 4–8
- Perennial
- 12–15" (30–38 cm)
- Filtered shade; protection from hot, summer sun
- Moderate

Harris checkerspot butterfly

220

New England Aster
Aster novae-angliae

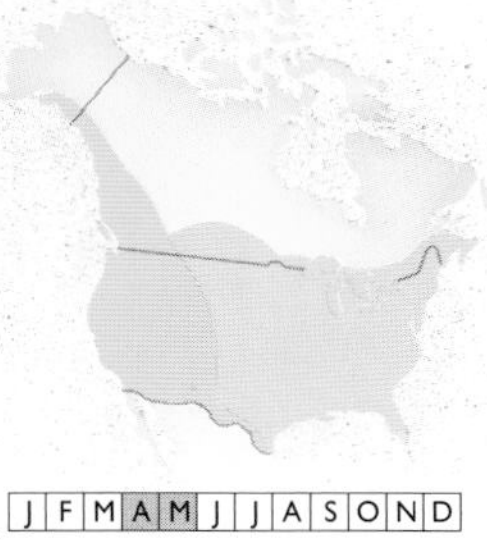

The New England aster grows in moist soil, and is often found along streams and roadsides, where its rich purple or red–violet ray flowers accent the tawny reds and yellows of the fall landscape. The plant may reach 6' (1.8 m) in the wild but it can be kept to a more compact size by shearing or mowing in early summer. While it prefers a moist position in the garden, it will also do well in an average perennial bed.

A stately plant, its narrow, downy, light green leaves clasp the stem at the base, and showy, daisy-like flowers appear in late summer to fall. It is adaptable, easy to divide and move, and grows readily from seed.

J F M A M J J A S O N D

common sulfur butterfly

All asters attract butterflies, as they are among the few nectar sources available so late in the season. Among those that visit New England aster are painted ladies, tiger swallowtails, common sulfurs, and monarchs. American toads are common throughout the Northeast, particularly in moist areas where asters grow.

FIELD NOTES
- Zones 4–9
- Perennial
- 6' (1.8 m)
- ☀ Full sun
- ◊ Moderate

American toad

221

Bee Balm

Monarda didyma

In midsummer, bee balm's ragged, scarlet pompoms bloom in low meadows and along the sunny banks of streams. It is a plant that grows well in any garden bed but really comes into its own in a moist meadow. It spreads readily in rich, moist soil and is, therefore, both easier to control and far more effective when given plenty of room.

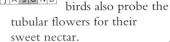

Bee balm is a plant for full sun—it tends to flop in the shade. If you remove the spent flowers, the plant will go on blooming for several weeks. Easily divided, bee balm is also readily grown from seed, although there may be considerable color variation in the resulting plants. In areas of poor air circulation, it may be susceptible to mildew, especially where the summers are hot and humid,

honeybee

but there are mildew-resistant plants among the many available selections. A blooming patch of bee balm will attract bees and clouds of colorful butterflies, including monarchs, viceroys, painted ladies, and fritillaries. Quick, darting, ruby-throated humming-birds also probe the tubular flowers for their sweet nectar.

ruby-throated hummingbird
♂

FIELD NOTES

■ Zones 4–9
■ Perennial
■ Deciduous
■ 3–4' (1–1.2 m)
☀ Full sun
◐ Moderate–high

222

Black-eyed Susan

Rudbeckia hirta

Swaths of black-eyed Susans, that familiar yellow daisy with the dark brown center, color our eastern meadows bright gold every summer. Yet, like many of our meadow flowers, the fast-growing black-eyed Susan originated elsewhere: it is a mid-western Prairie flower that spread eastward when the forest was cleared.

Rudbeckia needs full sun but will grow in any ordinary soil. It is grown from seed. A short-lived perennial, it must continually reseed itself or eventually die out. There are many cultivars on the market; the widely advertised gloriosa daisies are nothing more than tetraploid black-eyed Susans.

J F M A M J J A S O N D

The black-eyed Susan, like all members of the daisy family, is a composite flower. What we see as a single yellow and brown blossom is actually a ring of tubular ray flowers surrounding a tight cluster of flowers. All composites attract butterflies; they provide ample room for landing and perching as well as numerous sources of nectar. In addition, the black-eyed Susan is a food plant for the larvae of a small orange-and-black butterfly called the silvery checkerspot.

monarch butterfly

FIELD NOTES
- Zones 5–8
- Annual/biennial/perennial
- 3' (1 m)
- Full sun
- Low–moderate

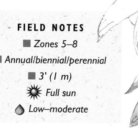

223

■ Woolly blue violet (*Viola sororia*)

Violets

Viola *spp.*

There are no fewer than 52 species of violet in the Northeast. Colors range from white, to yellow, to blue, to deep purple. There are stemmed violets and stemless violets. Most are woodland plants, but some grow in meadows, and one, the pansy-like bird's foot violet (*V. pedata*), prefers poor, sandy soil. Violets have leaves in a variety of shapes. All, however, have five-petaled flowers that are bilaterally symmetrical; that is, the halves are equivalent only when the flower is bisected vertically.

Woodland violets thrive in rich soil under high trees or at the edge of woods. They are easily propagated from seed, and those with runners or multiple crowns can be divided. Some species spread widely on their own, producing quantities of seed in unique, closed, self-fertilizing flowers.

A group of butterflies called fritillaries uses violets exclusively as host plants: it includes the great spangled, Aphrodite, Atlantis, meadow, and silver-bordered fritillaries. The regal fritillary feeds on bird's foot violets. Spring azure butterflies are also attracted to the nectar of these plants.

| J | F | M | A | M | J | J | A | S | O | N | D |

spring azure butterfly

FIELD NOTES
■ *Zones vary with species; most 4–10*
■ *Perennial*
■ *1' (30 cm)*
☀ *Full sun or filtered shade*
◊ *Moderate*

Bearberry

Arctostaphylos uva-ursi

Bearberry is a trailing shrub with small, leathery leaves on long, flexible branches. On sandy shores, rocky slopes, and ledges, a single bearberry plant may form a dense carpet, several yards wide, but less than 6" (15 cm) high. Clusters of small, white, urn-shaped flowers, tinged with pink at the lip, appear on dark red stems in May. The dark red fruits that follow begin to ripen in July and last into November. Bearberry's leaves are evergreen, turning an attractive bronze with the arrival of cold weather. The bark is papery and peels easily. A particularly attractive ground-covering shrub for coastal gardens, bearberry also thrives in inland locations, if its rather exacting requirement for dry, acid, sandy soil can be accommodated. Container-grown plants can be planted in spring and fall and propagation is by layering.

The fruits are a favorite not only of bears, as the name suggests, but also of deer, ruffed grouse, and songbirds. The hoary elfin butterfly lays its eggs on the foliage.

J F M A M J J A S O N D

summer fruits

FIELD NOTES

- Zones 2–10
- Evergreen
- Under 6" (15 cm) high but several yards wide
- ☀ Full sun
- 💧 Low–moderate

spring flower

Red-osier Dogwood

Cornus sericea (stolonifera)

Red-osier dogwood is a common shrub in New England, where it typically grows along the banks of streams. In winter, its bright red stems appear to glow against the snow. Readily adaptable to the garden, it is particularly attractive in late winter and early spring, when the color of the stems seems to intensify.

Clusters of small, creamy white flowers appear in late spring, and the white fruits ripen during July and August. The plant sends up shoots along the roots, eventually forming a sizeable clump. The twigs root readily, making it easy to propagate. Red-osier dogwood prefers a moist, acid soil. A water source or mulch can help to retain moisture. Combine it with paper birch against a background of evergreens to make an attractive winter planting for a northern garden.

All dogwoods are good wildlife-attracting plants. Among the many kinds of animal and bird that feed on the fruit of red-osier dogwood are rabbits, wood ducks, evening grosbeaks, robins, wood thrushes, and cedar waxwings. The larvae of the small, blue spring azure butterfly also feed on its leaves.

J F M A M J J A S O N D

American robin ♂

FIELD NOTES
- Zones 4–9
- Deciduous
- 10' (3 m)
- ☀ Full sun or partial shade
- ◊ Moderate–high

Winterberry

Ilex verticillata

Winterberry is a deciduous holly and, like all members of that group, it bears male and female flowers on separate plants. Bees readily pollinate the flowers, but there must be a male plant nearby for the female plant to set fruit. Winterberry is primarily a plant of the coastal plain, growing naturally on stream banks, in swamps, and at the edge of salt marshes. It is the perfect choice for low, poorly drained areas, but is adaptable to drier conditions. Propagation is by stem cuttings or layering.

In the wild, winterberry is a tall, upright shrub but smaller cultivars are available. It is a shrub that is inconspicuous in spring and summer, but comes into its own in fall and winter, when the medium green foliage turns yellow and the pea-sized, scarlet berries begin to ripen. They will cover the bare, black branches until January to February, enlivening the winter garden and feeding a wide range of songbirds, including mockingbirds, robins, and cedar waxwings. The woolly bear caterpillar feeds on winterberry leaves.

summer foliage

J F M A M J J A S O N D

FIELD NOTES
- ■ Zones 4–8
- ■ Deciduous
- ■ 10' (3 m) tall
- ☀ Half sun–full sun
- ◗ Moderate–high

woolly bear caterpillar

227

Spicebush

Lindera benzoin

In very early spring, the fragrant flowers of the spicebush appear as a haze of soft yellow in the still leafless woods and are immediately visited by the first bees of the season. The spicebush bears male and female flowers on separate plants. The bark is a rich, reddish brown, dotted on the branches with conspicuous white spots; both twigs and foliage emit a spicy, aromatic odor when broken.

J F M A M J J A S O N D

In the wild, this shrub is commonly found on the flood plains of woodland streams. In the garden, it is a tidy plant, always attractive yet never intrusive, easily grown and transplanted. Propagation is by cuttings. Spicebush prefers a moist, acid soil but does not require a wet location in the garden.

In fall, the smooth, green leaves turn a bright golden yellow and shiny red fruits appear on the female plants. These fruits are valuable food for wood thrushes and veeries before their long fall migration. They are also relished by quail, ruffed grouse, and pheasants. Rabbits and white-tailed deer browse on spicebush twigs in the winter, and the shrub is also one of only two known hosts of the beautiful spicebush swallowtail butterfly (the other is sassafras, *Sassafras albidum*).

spring flower

FIELD NOTES
- Zones 4–8
- Deciduous
- 3–10' (1–3 m) tall
- Filtered sun or partial shade
- Moderate–high

spicebush swallowtail butterfly

Bayberry

Myrica pensylvanica

B ayberry's glossy, olive-green leaves and graceful, mounded shape are attractive all summer and especially so in fall and winter, when the foliage turns bronze and the angular, twiggy branches of the female plants are covered with thick clusters of metallic-gray berries.

Bayberry thrives in the dry, sandy soil of coastal dunes and lake shores. In the garden, it combines well with red cedar (*Juniperus virginiana*) for an attractive winter grouping. Because bayberry bears male and female flowers on separate plants, however, you will have to include at least one male plant for every 10 females. Since nurseries rarely differentiate between the two, it is best to select plants in fall when the female plants bear fruit. Propagation is from root cuttings.

J F M A M J J A S O N D

The tree swallow, after eating insects all summer, consumes quantities of bayberries along its fall coastal migration route to Florida and Central America. The uneaten fruit remains on the plant until early spring, when it is stripped off by returning orioles, bluebirds, and yellow-rumped warblers.

FIELD NOTES
- ■ Zones 2–7
- ■ Deciduous
- ■ 4' (1.2 m) tall
- ☀ Full sun
- ◌ Low

tree swallow

229

Sow-teat blackberry (*Rubus alleghenienesis*)

Blackberry, Raspberry

Rubus *spp.*

The genus *Rubus* contains many species that interbreed freely, but there are two general types: the blackberry and the raspberry. Raspberries can be black or red and usually ripen in July. When picked the berry is hollow, like a small cup, leaving its white core on the shrub. A blackberry is not hollow. It keeps its core when

JFMAMJJASOND

picked and ripens in August. Both groups have white, five-petaled flowers, three-part leaves, and prickly stems, with one exception: the purple-flowering raspberry (*R. odoratus*).

blackberry flower

The woody stems, called "canes", are short-lived, but the roots go on indefinitely. First-year stems are unbranched; they develop side branches which bloom and set fruit during the second summer, after which they die and can be removed. The canes can arch over, sometimes rooting where they touch the ground. For easier handling, keep them about 6" (15 cm) apart and cut back to 4' (1.2 m) high. Propagation is by stem or root cuttings.

Rubus berries are well loved by birds as well as foxes, bears, raccoons, and squirrels. The canes also provide birds with valuable nesting cover.

blue-winged warbler ♂

FIELD NOTES

■ Zones 4–8

■ Deciduous

■ 5–6' (1.5–1.8 m)

☀ Full sun

◆ Moderate

230

Highbush Blueberry

Vaccinium corymbosum

Highbush blueberry is the common, tall blueberry of swamps, bogs, and low wet ground, but it also invades old fields and pasture lands where the soil is considerably drier. In spring, the shrub is festooned with small, white or pink, bell-shaped flowers, followed in summer by the sweet, blue-black berries that we all love. The new twigs are yellow-green in summer, turning pale red in winter, and in fall, the smooth, green leaves turn a deep, brilliant red. While any of these attributes would be reason enough to grow highbush blueberry in the garden, older shrubs also acquire a twisted, gnarled trunk, reminiscent of ancient Japanese gardens. Highbush blueberry prefers a rich, moist, acid soil and can be propagated by cuttings or layering.

J F M A M J J A S O N D

Another attractive species is the lowbush blueberry (*V. angustifolium*). It is a northern shrub of open, rocky hillsides that produces quantities of sweet berries.

Blueberry flowers are pollinated by a variety of native bees. In the north, black bears are well-known for their fondness for the fruit, as are chipmunks and other small mammals. Scarlet tanagers, bluebirds, catbirds, and mockingbirds are but a few of the many songbirds that feast on them.

FIELD NOTES
- Zones 4–9
- Deciduous
- 10' (3 m)
- ☀ Full sun or partial shade
- ◌ Moderate–high

eastern chipmunk

231

Highbush Cranberry

Viburnum trilobum

Highbush cranberry is one of a large clan of native viburnums that make attractive garden plants. A tall, arching shrub, it is found growing at or near the edge of woodlands and swamps, and prefers a moist, rich soil. In May, highbush cranberry is laden with lacy, white flower clusters, each surrounded by a ring of large, sterile flowers. The maple-shaped leaves are opposite, like all members of the honeysuckle family, and turn red in fall. The shiny, red, translucent fruits ripen in fall and hang on for most of winter. Propagation is by seed or cuttings.

Highbush cranberry is a plant of the northern woods and may not thrive in hot, humid summers. Other viburnums enjoy those conditions, however, including black haw (*V. prunifolium*), withe-rod (*V. cassinoides*), and possum haw (*V. nudum*). These viburnums' fruits are spectacular; starting in late August, they change gradually from green to white to pink and finally to blue-black. The show may be brief, however, if red foxes, chipmunks, cedar waxwings, and white-tailed deer get to them first! Viburnums also host the larvae of two butterflies, the spring azure and Henry's elfin, and provide nectar for a broad range of others.

J F M A M J J A S O N D

Baltimore
butterfly

spring
flower

FIELD NOTES

■ Zones 4–6

■ 10' (3 m)

■ Deciduous

☀ Full sun; partial shade

💧 Moderate–high

232

Allegheny Serviceberry

Amelanchier laevis

Loose clusters of white Allegheny serviceberry blossoms appear in early spring, just as the shad swim up the cleaner eastern rivers to spawn. Each flower has five slender petals and numerous stamens. It is pollinated by bees. In fall, the foliage turns red or yellow and the light gray bark is attractive all winter.

The many eastern species of serviceberry tend to hybridize freely, making identification difficult. Most, however, are shrubs. Allegheny serviceberry is a small tree, one of only two to reach such a height. The other is downy serviceberry (*A. canadensis*). Allegheny serviceberry thrives in moist, acid soil. Propagation is difficult but plants are readily available from nurseries. A member of the rose family, it may be susceptible to insect damage in some areas.

J F M A M J J A S O N D

The red fruits ripen in June, providing food for wildlife at a time when few other fruits are ripe. In the wild, these early summer fruits are relished by everything from bears to wild turkeys, and they will attract many songbirds, including bluebirds, cardinals, evening grosbeaks, orioles, and tanagers, to the garden.

scarlet
tanager
♂

FIELD NOTES
- ■ Zones 3–8
- ■ Deciduous
- ■ 25' (7.6 m)
- ☀ Partial shade preferred
- 💧 Moderate–high

233

Paper Birch

Betula papyrifera

A graceful, often multitrunked tree from the far north, paper birch's distinctive, white, peeling bark once covered canoes and wigwams of Native Americans. The bark is a smooth, reddish brown on young trees, but it turns creamy white with age and peels readily. In the garden, the clean, white bark shows to best advantage against a background of evergreens.

J F M A M J J A S O N D

gray comma
anglewing
butterfly

The pointed, dark green leaves turn a clear yellow in fall.

Although widely planted throughout the East, paper birch is less susceptible to insects and disease where summers are cool. Even in the North, it usually occurs on north- or east-facing slopes, and farther south, is found only in the mountains. It is relatively fast-growing and prefers a well-drained, sandy loam. Paper birch is readily available from nurseries.

Paper birch seeds provide valuable food for birds such as the redpoll and pine siskin. The larvae of the gray comma, an anglewing butterfly, feed on the foliage. So do the Compton tortoiseshell, the mourning cloak, and the Canadian tiger swallowtail.

FIELD NOTES
- Zones 2–5
- Deciduous
- 80' (24 m)
- ☀ Full sun (north); partial shade (south)
- ◗ Moderate

Eastern Redbud

Cercis canadensis

Clusters of bright, purple-pink, pea-like blossoms appear along the bare branches of the eastern redbud in May. It looks lovely planted in combination with flowering dogwood (*Cornus florida*).

fall leaves

Eastern redbud occurs naturally inland as far north as southern Michigan. Farther east, it is confined to the piedmont, south of Pennsylvania, yet it will succeed in gardens north to southern New Hampshire and Vermont.

A compact tree, it has a short trunk and smooth, gray bark that develops scaly plates as the tree ages. The usual flower color is a bright purple-pink, but there are selected slight variations. The leaves are heart-shaped and have a papery texture. Bright green in summer, they turn yellow in fall. The seeds are borne in pods.

J F M A M J J A S O N D

Eastern redbud blooms well in filtered shade. In full sun it prefers soil which is rich and moist. Propagation is slow. It is valued more as a garden ornamental than a wildlife tree, but it does provide larval food for the small, brown Henry's elfin butterfly, and provides good nesting branches for blue jays and other birds.

blue jay

♂

FIELD NOTES

- Zones 5–10
- Deciduous
- 25' (7.6 m)
- ☀ Full sun or partial shade
- ◊ Moderate

235

American Beech

Fagus grandifolia

A mature American beech is a majestic sight. In the woods, it grows straight toward the light, forming a broad crown in the forest canopy. When grown in the open, however, its broad, welcoming branches frequently sweep the ground.

The bark is a smooth, light gray, often with horizontal folds at the base, giving the trunk the look of an elephant's foot. Long, pointed buds unfold in early spring, and the leaves often remain through winter, in fall turning the light brown of old parchment and rustling softly in the breeze. Both male and female flowers appear on the same tree.

fall foliage

J F M A M J J A S O N D

The American beech grows best in rich, moist loam. It is found in moist forests throughout the East, reaching its greatest size in the Mississippi and Ohio river valleys. Though easily grown from seed, the tree's slow growth suggests planting from nursery stock is preferable.

Triangular nuts, contained in a spiny husk, ripen by the first frost. Black bears and porcupines particularly depend on them; chipmunks, squirrels, chickadees, and tufted titmice all love them. The larvae of the early hairstreak butterfly also feed on the leaves.

fall husk and beech nuts

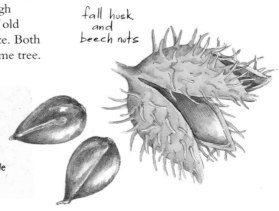

FIELD NOTES
- Zones 3–8
- Deciduous
- 100' (30 m)
- Full sun or shade
- Moderate

236

Eastern White Pine

Pinus strobus

Widely planted throughout the East, eastern white pine has soft, dense, feathery, blue-green needles in bundles of five, and an attractive pyramidal shape when young. As the tree ages, it tends to drop some of its branches, eventually acquiring an irregular silhouette and sweeping, horizontal crown.

When considering an eastern white pine for the garden, remember that this is a fast-growing tree eventually reaching a great size and needing ample room to do so. You will need to be watchful for white pine blister rust and white pine weevil. It prefers a well-drained soil. Propagation is by seed, but it is readily available from nurseries.

A mature eastern white pine is not only one of the most conspicuous and beautiful components of the eastern landscape, but also a good food source for wildlife. Red-breasted nuthatches, black-capped chickadees, red crossbills, squirrels, chipmunks, and mice, feed on the seeds in the pendulous cones which take two years to mature. Beavers, porcupines, and white-tailed deer browse on the foliage and twigs.

J F M A M J J A S O N D

black-capped chickadee ♂

FIELD NOTES

- Zones 3–6
- Evergreen
- 150' (45 m)
- ☀ Full sun
- ◐ Moderate

Pin Cherry

Prunus pensylvanica

In much of the East, pin cherry is a small, early-blooming tree of fields and pastures. Farther north, it is known as fire cherry, because of its propensity to move into a forest after a fire, providing shade for the seedlings of the trees that will eventually replace it. It also grows plentifully along the banks of prairie rivers.

white admiral butterfly

J F M A M J J A S O N D

In the garden, pin cherry is a medium-sized tree with a rounded crown. It is beautiful in early spring when clusters of small, white flowers cover the branches, and again in late summer when the red fruits are ripe. Green summer foliage turns reddish, then bright yellow, in fall. Like other wild cherries, it may be susceptible to disfiguring attacks by the tent caterpillar in early summer. Propagation is by seed or cuttings, but it is faster to purchase plants from the nursery. Pin cherry is adaptable to a variety of soils.

Over 30 birds in the Northeast eat the cherries. They are relished by grouse and pheasant, as well as evening and rose-breasted grosbeaks, bluebirds, and thrushes. Flocks of robins fly back and forth to cherry trees, picking off the fruit. Chipmunks, squirrels, and foxes feed on fallen fruit, and the bark and twigs attract browsing deer. Pin cherry also serves as a host plant for the larvae of coral hairstreak, spring azure, and white admiral butterflies.

FIELD NOTES
■ Zones 3–6
■ Deciduous
■ 40' (12 m)
☀ Full sun
◌ Moderate

238

White Oak

Quercus alba

The regal white oak is truly the king of the eastern forest; it is a tree that grows tall and straight in the woods, but spreads its mighty branches in more open situations. The leaves are smaller than those of many oaks, with deeply cut, rounded lobes, and the acorns ripen in a single season.

The white oak's majestic size and spreading crown bring a sense of maturity and permanence to any garden. In spring, the unfolding leaves are a velvety pink, and in fall, the whole tree turns burnt orange, deep rich carmine red, or rich brown. Oaks are always among the last to turn, and often retain their leaves into early November. Young trees may keep them all winter, providing cover for birds and other animals. White oak grows best in rich, moist, well-drained loam. Propagation is from seed but is very slow.

J F M A M J J A S O N D

White oak acorns, like those of all oaks, provide regular food for wood ducks, wild turkeys, blue jays, nuthatches, titmice, chickadees, woodpeckers, bears, raccoons, mice, and white-tailed deer. Squirrels often bury the acorns for later use, inadvertently planting more oaks in the process. Returning migrants, such as warblers, vireos, and tanagers, also flock to oak trees for the insects the trees harbor in spring.

new spring leaf

gray squirrel

FIELD NOTES

- Zones 4–8
- Deciduous
- 115' (35 m)
- Filtered sun
- Low–moderate

239

Sassafras

Sassafras albidum

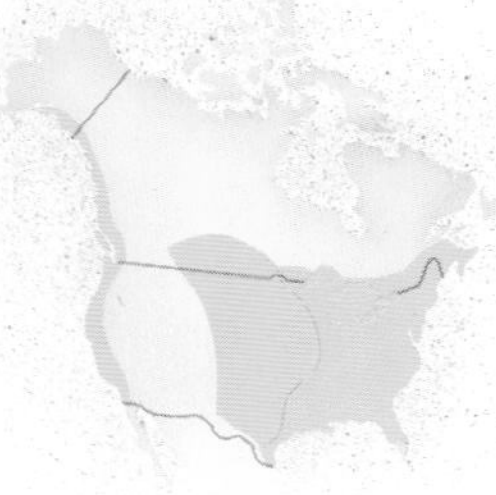

Sassafras is truly a tree for all seasons. It has bright, greenish-yellow flowers in early spring, brilliant red-orange leaves in fall, green twigs in winter, and an attractive, horizontal branching pattern. It is one of the first trees to move into old fields and is also common in hedgerows throughout the East, where it grows from seeds dropped by birds that eat the fruit.

As a garden tree, sassafras is both underused and underappreciated. Difficult to propagate, it is rarely seen in nurseries, possibly because large specimens can be difficult to transplant. Sassafras spreads by underground runners. In the wild, a single sassafras tree will soon sprout a ring of satellite trees from its roots. This habit of growth can also be used attractively in the garden. Sassafras prefers a sandy, well-drained loam.

J F M A M J J A S O N D

The deep blue fruit is never plentiful, but is eaten by a number of birds, including two that normally eat only insects: the eastern kingbird and the phoebe. Catbirds and pileated woodpeckers also love sassafras fruit and the tree is one of only two known hosts of the larvae of the spicebush swallowtail butterfly (the other is spicebush, *Lindera benzoin*).

FIELD NOTES
■ Zones 5–8
■ Deciduous
■ 60' (18 m)
☀ Full sun
💧 Moderate

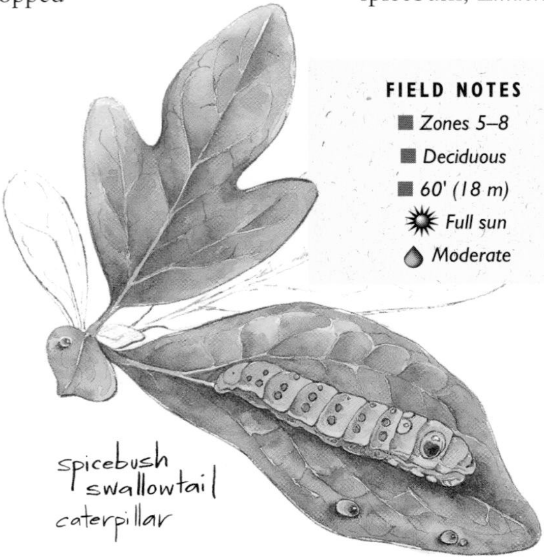

spicebush
swallowtail
caterpillar

Canadian Hemlock

Tsuga canadensis

In the wild, Canadian hemlock grows on cool slopes, rocky ridges and ravines, often in pure stands. It is a dark, dense evergreen with small, flat needles and slightly pendulous branches that sway gracefully in the wind. Each stalked needle is a lustrous dark green above and silvery white beneath. Perfect miniature cones hang from the tips of the branches. Canadian hemlock is both slow-growing and very long-lived; an individual tree may live to be several hundred years old. In the garden, hemlocks make good background plants and can be maintained at almost any height. For this reason, they make excellent informal hedges. They prefer a

J F M A M J J A S O N D

moist, acid soil. Canadian hemlock is widely available from nurseries. Hemlocks in the East are presently under attack by a prolific insect pest called the woolly adelgid. Insecticidal soap or oil (applied in July–September) is effective in treating infested garden trees.

Hemlocks are good wildlife trees. The dense, low foliage of young trees provides good winter cover for a number of birds and animals. Birds that seek out hemlocks for nesting sites are veeries, black-throated blue warblers, and northern juncos. The small, winged seeds feed pine siskins, crossbills, chickadees, and red squirrels.

hemlock cones

FIELD NOTES

■ Zones 4–6

■ Evergreen

■ 80' (24 m)

✺ Full sun–deep shade

◉ Moderate

red squirrel

241

Virginia Creeper

Parthenocissus quinquefolia

Virginia creeper is an easily grown vine which is among the first plants to change color in fall. As early as midsummer, the leaf petioles turn a brilliant red, contrasting dramatically with the dark green leaves and ripening blue-black fruit. The foliage gradually turns a deep red, eventually fading to a translucent pink.

The most conspicuous feature of Virginia creeper is its five-part leaf, which immediately distinguishes it from that other well-known eastern vine: poison ivy (*Rhus radicans*)! The vine climbs by means of adhesive disks. It can be grown up tree trunks without damaging them, and looks particularly attractive draped over stone walls.

A vigorous grower, yet easily kept in check by judicious pruning, Virginia creeper thrives in any kind of soil and does equally well in sun or shade. It will bloom and set fruit far more plentifully, however, if given a few hours per day of bright sunlight. Propagation is by seed or layering.

Bluebirds, mockingbirds, and woodpeckers, including the handsome pileated woodpecker, relish the fruit of this vine. Virginia creeper is the main food plant for the three eastern species of sphinx moth: the myron, Pandora, and white-lined sphinx moths.

J F M A M J J A S O N D

FIELD NOTES
- Zones 4–10
- Deciduous
- 30' (9 m)
- Sun or shade but prefers some direct sunlight
- Low–moderate

northern mockingbird ♂

242

Concord grape (Vitis labrusca 'Concord')

Wild Grape

Vitis spp.

grapevine beetle

Grapes are large-scale vines that cling by means of tendrils to the slender branches of trees. They have large, flat leaves which can smother young plants by blocking out light. The flowers of all grapes are inconspicuous, but intensely fragrant. The heart-shaped leaves often have white or tan felt on the underside. Their fruits range from tart to musky sweet.

J F M A M J J A S O N D

There are about 30 species of grapevine native to North America, many of which interbreed freely. Grapevines are fast-growing and can be readily trained on an arbor to shade a terrace. They prefer a rich, well-drained loam and can be propagated from cuttings. For best production of fruit, prune them back in early spring to a few buds on the previous year's wood. One native species that adapts well to the garden is the fox grape (*V. labrusca*).

A common pest of wild and cultivated grapes, found throughout the East, is the grapevine beetle.

Gold- and purple finches, catbirds, mockingbirds, brown thrashers, and cardinals commonly use strips of peeling grapevine bark in constructing their nests. Most also eat the fruit, as do robins, fox sparrows, cedar waxwings, pileated and red-bellied woodpeckers, foxes, black bears, and skunks. The larvae of various moths and butterflies feed on the leaves.

white-lined sphinx moth caterpillar

FIELD NOTES

■ Zones vary with species; most 4–10

■ Deciduous

■ 20' (6 m)

☀ Sun or partial shade

◊ Moderate

243

Turk's-cap lily (left); chokeberry (right)

Additional species

Plant name/flower color	Zone	Height	Visitors/attraction/season
HERBACEOUS FLOWERS			
Rock cress (white) *Arabis caucasica* (exotic)	4–9	9" (22 cm)	butterflies/flowers/spring
False spiraea (various) *Astilbe* spp. (exotic)	4–8	2–4' (60–120 cm)	butterflies/flowers/spring–summer
Turtlehead (pink, white) *Chelone glabra*	4–9	1–3' (30–90 cm)	caterpillars/leaves/summer
Bunchberry (white) *Cornus canadensis*	2–5	3–8" (7.5–20 cm)	birds/fruits/fall
Coralbells (red) *Heuchera sanguinea*	3–10	2–3' (60–90 cm)	hummingbirds/flowers/spring–summer
Rose mallow (pink) *Hibiscus moscheutos palustris*	6–9	4–5' (120–150 cm)	butterflies/flowers/summer
Turk's-cap lily (orange) *Lilium superbum*	4–8	3–6' (90–180 cm)	hummingbirds/flowers/summer
Summer phlox (pink, white, lavender) *Phlox paniculata* (exotic)	4–8	3' (90 cm)	hummingbirds, moths, butterflies/flowers/summer
SHRUBS			
Chokeberry (white) *Aronia arbutifolia*	4–9	3–8' (90–240 cm)	caterpillars/leaves/spring birds, foxes/fruits/fall–winter
Buttonbush (white) *Cephalanthus occidentalis*	4–8	3–10' (90–300 cm)	butterflies, bees/flowers/summer ducks, deer/seedheads/fall
Flowering quince (red, pink, white) *Chaenomeles japonica* (exotic)	4–8	6–10' (1.8–3 m)	hummingbirds/flowers/spring birds/nesting/spring–summer
Witch hazel (yellow) *Hamamelis virginiana*	4–8	12' (3.6 m)	grouse, squirrels/seeds/fall–winter
Mountain laurel (pink, white) *Kalmia latifolia*	4–8	3–12' (90–360 cm)	ruffed grouse, deer/leaves/all year
Sargent crab apple (pink, white) *Malus sargentii* (exotic)	4–8	6' (1.8 m)	birds/fruits/fall butterflies, bees/flowers/spring
Shining sumac (yellow) *Rhus coppallina*	5–9	2–8' (60–240 cm)	rabbits, deer, birds/fruits/fall
Elderberry (white) *Sambucus canadensis*	4–9	5–8' (1.5–2.4 m)	birds/fruits/summer
TREES			
Sugar maple (red-orange foliage) *Acer saccharum*	3–8	90' (27.5 m)	birds, chipmunks, squirrels/ seeds/summer
American hornbeam (green) *Carpinus caroliniana*	3–9	20–30' (6–9 m)	birds/nuts/fall birds/nesting/spring–summer
Willows (yellow, white) *Salix* spp.	2–10	10–40' (3–12 m)	caterpillars/leaves/spring–summer birds: nest/spring–summer; buds/spring
American mountain ash (white) *Sorbus americana*	3–8	10–30' (3–9 m)	birds/fruits/fall

The Southeast

THE SOUTHEAST
The Woodland Garden

In the Southeast region, from Florida to eastern Texas and north to Tennessee and Virginia, gardeners must deal with extreme heat and humidity for much of the year. They are fortunate, however, to have a large and diverse group of native trees and shrubs to choose from in creating a cool, shady woodland garden that is a haven not only for mammals, butterflies, and birds, but for people, too. Plants such as the deciduous fringe tree (*Chionanthus virginicus*) with its fleecy, fragrant, white flowers, the lemon-scented sweet bay (*Magnolia virginiana*) with waxy, white flowers amidst evergreen leaves, and the red maple (*Acer rubrum*) with its brilliant fall foliage all help to render the natural woodland garden a place of year-round beauty.

Trees form the framework for the garden and absorb the sun's heat. Lower evergreen trees and large shrubs work well as screening for privacy and help to channel any cooling breezes through the garden. A woodland garden with both evergreen and deciduous trees—eastern red cedar (*Juniperus virginiana*) on the sunny margin, American holly (*Ilex opaca*) and flowering dogwood (*Cornus florida*) within the shelter of the woods—combines attractive and colorful flowers, fruit, and foliage. These same plants are a source of food and shelter for gray squirrels, bluebirds, catbirds, flickers, and many other birds.

Shrubs in the woodland garden include those grown for their fragrant flowers, varied foliage, and colorful and edible fruits. The hoary azalea (*Rhododendron canescens*), is just one of many species of rhododendron native to the area that provide a nectar source for butterflies, such as gulf fritillaries and swallowtails. Hoary azalea flowers best along the sunnier margins of the woodland. Cardinals, robins, and cedar waxwings dine on the dark fruits of inkberry (*Ilex glabra*) and sour gum (*Nyssa sylvatica*), both found in moist woodland areas.

The sunlight enters the garden through an opening in the woodland canopy, encouraging a meadow-like planting of herbaceous flowers, most notably the bright yellow tickseed (*Coreopsis lanceolata*), and the thick-leaved phlox (*Phlox carolina*).

A quiet pond or gurgling stream provides water for wildlife as well as dependably moist soil for dramatic clumps of cardinal flower (*Lobelia cardinalis*), whose brilliant red flowers are irresistible to hummingbirds. On the edge of the woodland, the flowering stems of trumpet honeysuckle (*Lonicera sempervirens*) reach up toward the sunlight and are a food source for butterflies and hummingbirds.

Tickseed

Coreopsis lanceolata

Tickseed is an ideal perennial for hot, dry situations. A tall plant, it produces bright yellow with some orange, 2" (5 cm), daisy-like flowers from mid-spring until July. Cultivars are available in pale yellow and other shades of yellow-orange. This species has been used in hybridizing many of the double-flowering *Coreopsis* common in nurseries.

Propagation is by seed or division. Tickseed prefers a sandy, moist, well-drained soil. It tolerates drought well. The flowering season will be extended if early flowers are deadheaded; later flowers will still produce seed in late summer.

Plant several varieties of *Coreopsis* and you will have blooms in the garden from early spring until the first frost. For a bright combination, combine tickseed with bee balm (*Monarda didyma*) and summer phlox (*Phlox paniculata*), both of which attract butterflies.

A colorful addition for a naturalized meadow planting or by itself, the seeds of this easy-care plant attract birds such as juncos, buntings, sparrows, and finches.

J	F	M	A	M	J	J	A	S	O	N	D

American goldfinch

♂

FIELD NOTES

■ *Zones 6–9*

■ *Perennial*

■ *2–3' +*
(60–90 cm) +

☀ *Full sun*

◐ *Low–moderate*

Joe-pye Weed

Eupatorium purpureum

Joe-pye weed is a large, impressive perennial that brings a welcome splash of color to the late summer and fall garden. There may be from five to as many as nine of the late summer, purple flower heads (sometimes with some pink) packed together, the overall diameter of these flower heads together measuring 12–18" (30–45 cm). The coarse-textured, 8–12" (20–30 cm), dark green leaves are in whorls of three to five leaves each; if crushed, they give off a vanilla-like scent. Joe-pye weed is a bold plant for the back of a border, along a stream, or by a pond, where it will get the moisture and full sun it prefers.

Propagation is by seed, division, or cuttings. Under most conditions, Joe-pye weed will become a massive clump, best suited to a large garden.

Late season butterflies that seek out its nectar include the large wood nymph, spicebush swallowtail, and zebra swallowtail.

J F M A M J J A S O N D

FIELD NOTES

■ Zones 4–9

■ Perennial

■ 4–7'
(1.2–2 m)

☀ Full sun–part shade

◯ High

*large
wood nymph
butterfly*

249

Cardinal Flower

Lobelia cardinalis

Cardinal flower is a beacon in the late summer (August) garden, with its tall spikes of bright red flowers and dark green leaves. It is ideal for a flower border with other perennials, in a woodland shade garden with ferns and other plants, or for a spot of color along a stream, as it might be found in the wild.

Propagation is by seed, offshoots, or stem cuttings. Cardinal flower prefers a moist, rich soil and tolerates both light shade and full sun, especially if kept moist. It resents extreme dry conditions. Blooming for several weeks, this plant is sure to attract the ruby-throated hummingbird if it's in the neighborhood, as well as the spicebush, two-tailed, and pipevine swallowtail butterflies. The green tree frog feeds on insects attracted to cardinal flower's blooms.

J F M A M J J A S O N D

green tree frog

FIELD NOTES
■ Zones 2–9
■ Perennial
■ 2–4'
(60–120 cm)
☀ Shade–part sun
◐ High

two-tailed swallowtail butterfly

250

Thick-leaved Phlox

Phlox carolina

Thick-leaved phlox is found growing in and along the edges of deciduous woodlands. The pink-purple flowers appear in big clusters on 3' (1 m)-tall stems, set off by thick, shiny green leaves in May to July. By pinching back plants in early spring, before they set flower buds, plants will be bushier and produce more flowers. Easy to grow, thick-leaved phlox does best with some shade but will take full sun. Feature it in flower borders, or place drifts at the edge of a shrub border or woodland garden.

Goldenrods (*Solidago* spp.) and asters (*Aster* spp.) are good companion plants for thick-leaved phlox.

Propagation is by seed or cuttings, and thick-leaved phlox prefers a moist, well-drained soil.

Butterflies that are attracted to the flowers include the eastern black swallowtail, the gray hairstreak, and the palamedes swallowtail.

| J | F | M | A | M | J | J | A | S | O | N | D |

gray hairstreak butterfly

FIELD NOTES
- ■ Zones 7–9
- ■ Perennial
- ■ 2–3' (60–90 cm)
- ☀ Shade–part sun
- ◊ Moderate

251

Red Buckeye

Aesculus pavia

Red buckeye grows as a large, clump-forming shrub or a small tree. Its showy flowers, a blend of red and yellow, brighten the woodland or shrub border in spring, especially when planted in masses. Red buckeye is also a good understory plant. The handsome, lustrous, dark green foliage has five to seven leaflets, and provides visual interest even when the plants are not flowering.

J F M A M J J A S O N D

fall fruits

spring flower

Propagation is by seed. The moist, fertile soils of woodland conditions are ideal for growing red buckeye; in such shaded conditions, red buckeye will be taller and more open than in full sun. Under stressful conditions, it should be mulched to reduce moisture loss.

The shiny seeds, which ripen in October, are thought to be a good luck charm by people who carry them, but are poisonous to cattle and to people if ingested. Ruby-throated hummingbirds are attracted to the flower nectar in spring.

FIELD NOTES

■ *Zones 4–8*

■ *Deciduous*

■ *10–20' (3–6 m)*

☀ *Full sun best, but can take dense shade*

💧 *Moderate*

Fringe Tree

Chionanthus virginicus

Fringe tree is an old-fashioned favorite, also called grancy gray-beard for its fleecy, beard-like, white flowers that perfume the air in spring. The dark green, leathery leaves turn a clear yellow in fall. A slow-growing, graceful shrub or small tree, it is ideal for gardens with only limited space, as a substitute for the flowering dogwood (*Cornus florida*). Plant fringe tree as a small specimen, or stand-alone, tree or as part of a shrub border along a woodland edge or stream.

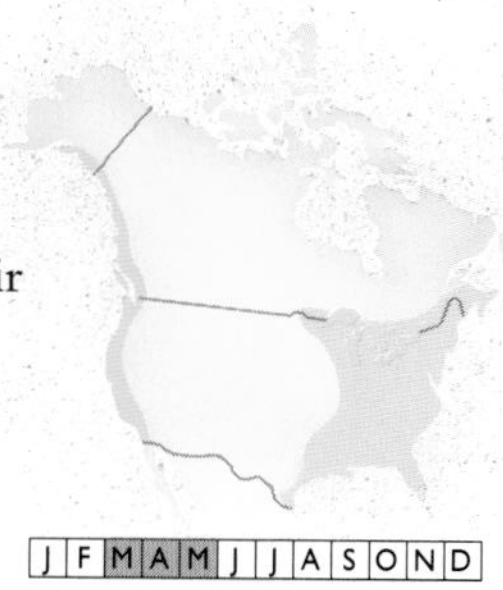

Propagation is by seed. Fringe tree likes a moist, rich, acid soil, but it adapts to a range of conditions. It should be watered during dry periods.

The dark blue, egg-shaped fruits that appear in late summer are favored by many birds, including bobwhites, gray catbirds, mockingbirds, robins, and starlings. Many birds also find suitable nesting sites among the fringe tree's branches.

| J | F | M | A | M | J | J | A | S | O | N | D |

FIELD NOTES

- Zones 3–9
- Deciduous
- 12–20' (3.6–6 m) or taller in the wild
- Full sun–part shade
- Moderate

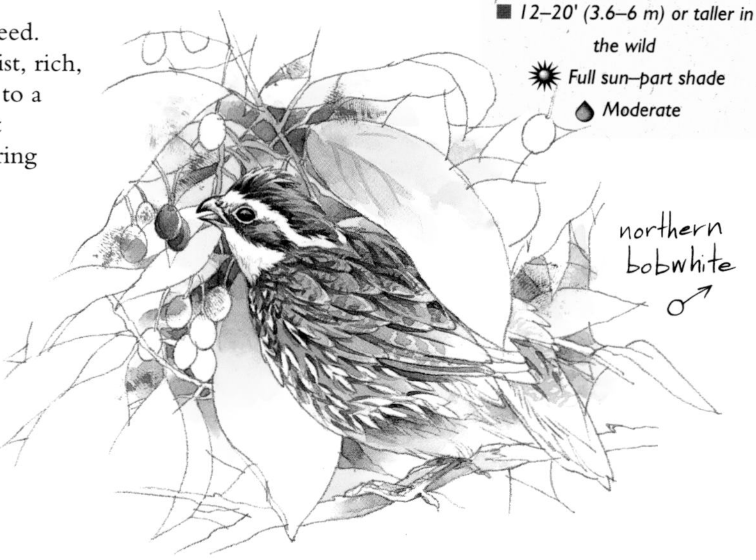

northern
bobwhite
♂

253

Ilex glabra 'Compacta'

Inkberry

Ilex glabra

Inkberry is a spineless, evergreen holly that is effective for hedges, screening, or in combination with deciduous shrubs and trees. The ink-colored, berry-like fruits last from September through to the following May. The small leaves, ¾–2" (2–5 cm) long and ⅓–⅝" (8–15 mm) wide, are a lustrous dark green, providing year-round color, as well as shelter and food for birds. Creamy flowers appear in late spring but are not showy; male and female flowers are produced on different plants. This adaptable shrub can grow to 8' (2.4 m) or more depending on the conditions.

J F M A M J J A S O N D

Easy to grow, it tolerates shade or sun and wet or dry soil, although moist, fertile soil is best. Propagation is by cuttings. Inkberry sometimes suckers and forms colonies. Be sure to plant at least one male and one female to be assured of some fruit production.

Birds that are attracted to the fruit of inkberry and other hollies include bluebirds, bobwhites, northern cardinals, gray catbirds, cedar waxwings, yellow-shafted flickers, blue jays, robins, and woodpeckers.

spring flowers

gray catbird ♂

FIELD NOTES
- ◼ Zones 4–9
- ◼ Evergreen
- ◼ 6–8' by 8–10' (1.8–2.4 m by 2.4–3 m)
- ☀ Full sun–part shade
- 💧 Moderate–high

254

Hoary Azalea

Rhododendron canescens

Hoary azalea, also called piedmont or Florida pinxter, is a large, upright shrub. Its fragrant, funnel-shaped flowers vary from white to pink to rose and appear from March to mid–April, often before the light green leaves have developed. Native along streams, hoary azalea is a wonderful addition to the sunny edge of a woodland planting. Easy to grow, it spreads readily by underground stems.

Hoary azalea prefers a moist, well-drained soil, but it will tolerate less than perfect conditions, including poor soils. Water it during drought conditions. Propagation is by seed or from cuttings.

Butterflies that love the sweet nectar of this native include the tiger and spicebush swallowtails and the American

| J | F | M | A | M | J | J | A | S | O | N | D |

painted lady. The leaves of the hoary azalea are also a source of larval food for the gray comma and the striped hairstreak butterflies.

gray comma butterfly

FIELD NOTES

- Zones 5–9
- Deciduous
- 10–15' (3–4.5 m)
- ☀ Full sun–part shade
- ◌ Moderate–high

255

Red Maple

Acer rubrum

Also called swamp maple, red maple thrives in flood plains and low-lying woodland areas. Always a beautiful tree, it is best known for its crimson fall foliage that appears to glow in the landscape. The leaves, 2–4" (5–10 cm) long and almost as wide, with three to five lobes, are attached to long, red stalks. They are medium green on top and gray underneath. The scarlet or yellow flowers appear in early spring before the leaves, making a show against bare, gray stems; in winter, the gray bark stands out against blue skies. Winged fruits called samaras appear a month after flowering and are a favorite of children, who like to watch them twirl through the air like miniature helicopters.

Propagation is from mature seed in early summer, which will germinate without special treatment, but germination can be hastened with stratification. Red maple also roots readily from softwood cuttings. It is adaptable to a wide range of soil types, including

swamp conditions, but prefers acid, moist soil, so ensure that it receives plenty of water. Plant the red maple as a shade tree, or as part of a mixed planting of trees and shrubs.

Birds that favor red maple for their nesting site include American robins, bobwhites, doves, and American goldfinches.

J F M A M J J A S O N D

FIELD NOTES

■ Zones 3–9

■ For best results, select trees grown from seed collected in the local area

■ Deciduous

■ 40–60' (12–18 m)

☀ Full sun–part shade

◗ Moderate–high

winged fruits
(samaras)

256

Pawpaw

Asimina triloba

Pawpaw is a small, multi-stemmed tree or shrub with huge, drooping leaves that create a dramatic effect, turning from a light green to a soft, butter yellow in fall, and often hiding the fruits. The curious flowers, which are 2" (5 cm) across, start out green and turn a dark maroon, before or as the leaves are developing, in late spring. Not everyone will like their strong scent, which is similar to that of fermenting fruit. The 3–5" (7.5–12.5 cm)-long fruits have a sweet aroma, but they are not ripe until they are wrinkled, ugly, and almost black, usually in October to November. Custard-like in texture, the fruit's taste has been compared to that of a banana/pear combination.

Propagation is from seed, but the seed and seed coats may exhibit some dormancy, making germination erratic. Pawpaw prefers a slightly acid, moist, fertile, and deep soil. It will tolerate some drought, especially if it is partially shaded.

The fragrant fruit is a favorite of opossums, raccoons, and gray squirrels. The pawpaw's leaves are a source of larval food for the zebra swallowtail butterfly.

| J | F | M | A | M | J | J | A | S | O | N | D |

late spring flower

zebra swallowtail butterfly

FIELD NOTES

- Zones 5–8
- Deciduous
- 15–20' (4.5–6 m) high and wide, under ideal conditions it may reach 30–40' (9–12 m)
- ☀ Full sun–part shade; in full shade, plants become leggy
- ◊ Moderate

Pignut Hickory

Carya glabra

Extensive oak–hickory woodlands still cover much of the eastern states; a major element of this woodland community is the pignut hickory, a large tree with compound leaves that turn a beautiful gold in fall. Early European settlers called it broom hickory, for the brooms they made from narrow strips of wood split from the tree. Flowers appear with the leaves in early spring, male flowers on catkins and female flowers on short, upright spikes; the fruits (pignuts) ripen in fall. As the tree matures, the smooth bark develops ridges, and the silhouette provides interest in the winter landscape.

fall fruits (pignuts)

Its long taproot makes the pignut hickory difficult to transplant, so start it from seed planted where the tree is to be grown. Be prepared to wait, as it may take 25 years or more to begin fruiting. Tolerant of a wide range of soil types and moisture levels, pignut hickory can be used by itself as a specimen, or stand-alone, tree or as part of a woodland planting.

In spite of their bitter taste, the nuts are popular with squirrels. The large trees provide nesting sites for blue grosbeaks, ruby-throated hummingbirds, blue jays, white-breasted nuthatches, and mockingbirds. The banded hairstreak butterfly larvae feed on the leaves.

J F M A M J J A S O N D

white-breasted nuthatch

FIELD NOTES
- ■ Zones 4–9
- ■ Deciduous
- ■ 50–60' (15–18 m), may grow as tall as 100' (30 m)
- ☀ Full sun–part shade
- 💧 Low

Sugarberry

Celtis laevigata

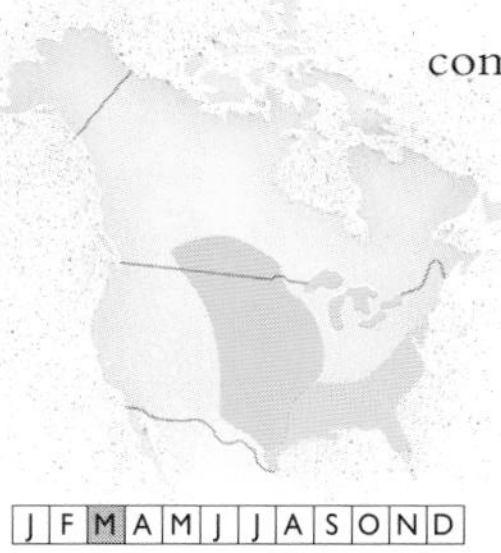

Sugarberry is a large, broad, rounded tree, sometimes with graceful, pendulous branches, that grows 60–80' (18–24 m) in height, with a similar spread. This adaptable tree tolerates compact, wet, clay soils. The overall effect of the simple, egg-shaped leaves—light green above, pale green below—is delicate. Tiny fall fruits change color from orange-red to dark purple.

Propagation is by seed, grafts, or cuttings. Plant sugarberry by itself in a large, open area or with smaller trees underneath it. Sugarberry is resistant to the witches' broom that

commonly disfigures other hackberries. It is named sugarberry because of its sweet, juicy fruit, which is a favorite of many birds, including bluebirds, blue jays, northern cardinals, mockingbirds, orioles, and robins. The sugarberry sap is a nectar source for the hackberry butterfly, and the young leaves provide important larval food for the snout, question mark, hackberry, tawny emperor, mourning cloak, and other species of butterfly.

J F M A M J J A S O N D

hackberry butterfly

FIELD NOTES

■ Zones 5–9

■ Deciduous

■ 60–80' (18–24 m)

☀ Full sun–part shade

💧 Moderate–high

Flowering Dogwood

Cornus florida

The flowering dogwood is an elegant native. A tree of year-round interest, it is perhaps best loved in spring, when its branches are covered with white flowers (bracts) blooming in the woods as well as in sunnier locations. A small to medium-sized tree, it grows wider than its height, with horizontal branches giving a layered effect that makes the tree stand out in the winter landscape. In fall, the dark green summer leaves turn red and purplish red.

J F M A M J J A S O N D

Glossy, red fruits ripen in September to October, until birds or other wildlife, such as the eastern gray squirrel, devour them.

Propagation is from seed and softwood cuttings. Flowering dogwood prefers acid, moist, and well-drained soil. Plant it in the garden as a single specimen, or stand-alone, tree or in groups with other dogwoods and native shrubs.

Flowering dogwood attracts a whole range of birds; who use the tree for cover, nesting, and as a source of food; including bobwhites, bluebirds, gray catbirds, mockingbirds, robins, sapsuckers, starlings, grackles, red-bellied woodpeckers, and quail. Young leaves are larval food for the spring azure butterfly.

spring flower

eastern bluebird ♂

FIELD NOTES

■ *Zones 5–9*

■ *Deciduous*

■ *20' (6 m), can grow 30–40' (9–12 m), wider than tall*

☀ *Full sun–part shade*

◌ *Moderate*

260

American Holly

Ilex opaca

American holly is a slow-growing, broadleaf evergreen associated with Christmas, when its red berries and dark green leaves are popular for use in wreaths and other decorations. Dense and pyramidal as a young tree, it becomes more open with age. Fruits mature in October, and often persist well into winter. To ensure that your holly will produce fruit, it is a good idea to plant both female and male of the same type; however, only the female will bear fruits.

Propagation is from cuttings, and American holly prefers an acid, well-drained soil. It does not tolerate heavy, wet soil or extreme, dry conditions. This tree is ideal for screening, hedging, as a specimen, or stand-alone, tree, or as a background for deciduous trees. It is easily pruned to control its size.

Birds that eat the fruit, or use the tree for shelter, include bluebirds, bobwhites, northern cardinals, gray catbirds, cedar waxwings, yellow-shafted flickers, blue jays, robins, and woodpeckers. Henry's elfin butterfly relies on American holly leaves for larval food.

J F M A M J J A S O N D

northern cardinal ♂

FIELD NOTES
- Zones 5–9
- Evergreen
- 30–50' (9–15 m); but 15–30' (4.5–9 m) is common in a garden
- Full sun–part shade, protect from drying winter winds
- Moderate

Eastern Red Cedar

Juniperus virginiana

Eastern red cedar is a large, tough, adaptable tree that grows across North America in a wide range of soils, in sunny, open places. It is perfect for hedgerows, screening, mass plantings, or mixed with deciduous trees. The habit varies from narrow, upright, and pyramidal to broad, open, and spreading. The cut wood and bruised leaves are intensely aromatic and the wood has long been popular for lining chests and closets.

J F M A M J J A S O N D

Flowering occurs in late winter with male and female flowers on separate trees. Female cones are greenish blue or violet; males are insignificant. Propagation is by seed or from cuttings and grafting. Eastern red cedar tolerates poor soils and high pH, but it prefers a moist, well-drained soil. Avoid using red cedar if there are apple trees nearby, as the two are alternate hosts of the cedar-apple rust, a serious disease of apples.

The sweet berries are the favorite of many birds, especially the cedar waxwing, which gets its name from this tree. Other birds that like red cedar fruits include bluebirds, gray catbirds, purple finches, mockingbirds, robins, yellow-bellied sapsuckers, flickers, starlings, grackles, and pileated woodpeckers.

cedar
waxwing
♂ ↗

FIELD NOTES

- ■ Zones 2–9
- ■ Evergreen
- ■ 40–50' (12–15 m) by 8–20' (2.4–6 m) wide
- ☀ Full sun
- 💧 Low

Tulip Tree

Liriodendron tulipifera

A stately tree, this member of the magnolia family is the tallest hardwood in North America, reaching heights of 150' (45 m). A large tree, it starts out pyramidal in form and develops an oval, rounded crown. The beautiful flowers, for which the tree is named, have six greenish-yellow petals with orange at the base. From April to early June they sit up above the leaves at the top of each branch. The foliage starts out as a glossy light green, turns deep green and then a golden yellow in fall.

Propagation is by seed or from cuttings, and it prefers a deep, moist, well-drained loam. Ensure that you water tulip

J F M A M J J A S O N D

trees during periods of drought. This is a choice species for use as a large specimen, or stand-alone, tree or in a large property; it is not suitable for the small garden.

Tulip tree is home to the eastern gray squirrel and birds such as northern cardinals and purple finches. Its leaves are a source of larval food for the tiger swallowtail and spicebush swallowtail butterflies. An excellent honey is produced by bees collecting the nectar.

spring flower

tiger swallowtail caterpillar

FIELD NOTES
- ■ Zones 4–9
- ■ Deciduous
- ■ 70–90' (21–27 m) tall, 35–50' (10.5–15 m) wide, can reach 150' (45 m) or higher
- ☀ Full sun
- 💧 Moderate

Sweet Bay

Magnolia virginiana

A small, evergreen to semi-evergreen tree, the dark green leaves with silver on the underside shimmer in the breeze, and the lemon-scented, creamy white flowers perfume the air from May to June, and then on and off until early fall. With its fragrant flowers, silvery leaves, and bright red, late summer to fall fruits, this small magnolia provides interest in the garden throughout the year, and makes a lovely specimen tree. It is often pruned as a shrub.

J F M A M J J A S O N D

This slow-growing, adaptable native can tolerate wet soils, making it ideal to plant along a stream bank or in a flood plain, by itself or in a mixed planting. It also tolerates acid, dry soil; propagation is by seed or from softwood cuttings.

Tiger swallowtail and spicebush swallowtail butterfly larvae eat the new leaves, and gray catbirds eat the seeds of the fruit.

seed head

spring-summer flower

FIELD NOTES
- Zones 5–9
- Evergreen
- 60' (18 m)
- Full sun–shade
- Moderate–high

264

Red Mulberry

Morus rubra

Red mulberry is an adaptable, fast-growing tree, best planted as part of a mixed planting of deciduous and evergreen trees. A large tree, it may be as wide as it is tall. It is not particularly ornamental, but many consider it more attractive than white mulberry (*M. alba*), which was used to feed silk worms to make silk. Red mulberry has darker green leaves that turn yellow in fall. Yellowish-green flowers appear in March to April. The approximately 2" (5 cm)-long fruits occur in June to July.

Propagation is by seed or cuttings, and red mulberry prefers a rich, moist soil. Red mulberry is not drought tolerant like white mulberry.

Birds and squirrels love the sweet, juicy fruits that resemble blackberries. Birds attracted to the red mulberry include bluebirds, bobwhites, purple finches, indigo buntings, doves, orioles, eastern kingbirds, mockingbirds, and robins.

J F M A M J J A S O N D

indigo bunting

♂

FIELD NOTES

■ Zones 5–9

■ Deciduous

■ 40–70' (12–21.5 m)

☀ Full sun–light shade

💧 Moderate

265

Sour Gum

Nyssa sylvatica

fall fruit & leaves

S our gum, or black gum, is an excellent native specimen, or stand-alone, tree or can be used as part of a woodland planting. Pyramidal as a young tree, its branches spread horizontally and it acquires a more rounded habit as it matures. It has beautiful, thick, shiny green foliage all summer, and in fall, it turns shades of brilliant red, orange, and gold. It is often one of the first trees to exhibit its fall color before other leaves have started to turn. A tree for all seasons, in winter, the checkered or scaly bark is picturesque in the landscape. Winter buds turn a brilliant red in spring.

J F M A M J J A S O N D

Propagation is by seed. Because of its taproot, sour gum is difficult to transplant. It is best to transplant balled and burlapped plants in early spring (March or April). Container-grown plants are also best planted in spring. Sour gum prefers moist, well-drained soil, but it is found in swamps and also in very dry conditions.

The dark blue fruits ripen in fall and are eaten by birds, such as bobwhites, painted buntings, crows, purple finches, yellow-shafted flickers, blue grosbeaks, and robins.

true katydid

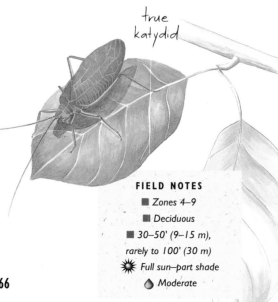

FIELD NOTES
- Zones 4–9
- Deciduous
- 30–50' (9–15 m), rarely to 100' (30 m)
- Full sun–part shade
- Moderate

266

Southern Live Oak

Quercus virginiana

Southern live oak is a large, majestic, oval-crowned tree. It is perfect as a specimen, or stand-alone, tree in a large garden. With a large, short trunk and massive horizontal and arching branches, often covered with Spanish moss, it is usually wider than it is tall with a spread of 60–100' (18–30 m). Sometimes, the branches reach down and touch the ground. Southern live oak's thick, narrow leaves, which start out olive green and turn a dark green, provide a fine-textured canopy. Inconspicuous, yellowish flowers appear in clusters in early spring.

J F M A M J J A S O N D

Southern live oak is adaptable to most soil types, including sandy, moist, and compact soils. This slow-growing tree will grow well in dry conditions and also tolerates salt spray. Propagation is by seed.

Squirrels and birds favor this oak for shelter and for its acorns. Birds that find southern live oak appealing include both common and colorful species, such as bobwhites, indigo buntings, crows, grackles, ruffed grouse, ruby-throated hummingbirds, and blue jays.

common grackle

♂

FIELD NOTES
- Zones 7–10
- Evergreen
- 40–80' (12–24 m)
- ☀ Full sun
- ◐ Moderate

Cabbage Palm

Sabal palmetto

For a tropical effect, the cabbage, or sabal, palm is the perfect tree. Its foliage resembles giant fans, each 3–5' (1–1.5 m) long by 2–3' (60–90 cm) wide. Growing in a wide range of soil types, from swampy to sandy, this palm looks good in groves, clumps, or as a single specimen, or stand-alone, tree.

J F M A M J J A S O N D

Large, drooping clusters of the flowers hang down below the leaves and the fruits are black and shiny. Young seedlings start out by growing downward into the soil before they turn upward permanently. A slow grower, it can reach 20–80' (6–24 m) by 6–10' (1.8–3 m) wide. Full sun or partial shade is best, and when transplanting, older plants do better because of the water they have stored for use during drought times. Propagation is by seed.

Birds that are attracted to the late summer fruits of cabbage palm include mockingbirds, robins, sapsuckers, warblers, fish crows, pileated woodpeckers, and red-bellied woodpeckers.

red-bellied woodpecker ♂

FIELD NOTES
- Zones 8–10
- Evergreen
- 20–80' (6–24 m)
- Full sun–part shade
- Low–moderate

268

Dutchman's Pipe

Aristolochia durior

Dutchman's pipe is an old-fashioned, vigorous, climbing, twining vine. Its huge, heart-shaped leaves, 4–8" (10–20 cm) across or larger, and its 3" (7.5 cm)-long flowers shaped like a pipe, make this curious vine a conversation piece. The spring flowers are yellow-green on the outside and brownish purple at the opening of the pipe. It needs support for climbing and does well when trained up a trellis, fence, or tree. The curious pollination strategy of this plant involves the flowers releasing a faint, unpleasant fragrance which draws flies and other minute insects into the pipe, where they become trapped. In their struggle to find an exit, they are pressed into contact with male and female parts, and pollination occurs.

spring flower

Propagation is by division or cuttings. Dutchman's pipe tolerates most soil types, as long as they are moist and well drained. It requires water during drought conditions.

The leaves of Dutchman's pipe are a source of larval food for the pipevine swallowtail butterfly.

J F M A M J J A S O N D

FIELD NOTES
- Zones 4–8
- Deciduous
- 20–30' (6–9 m) in a single season
- ☀ Full sun–part shade
- ◊ Moderate

pipevine swallowtail caterpillar

269

Trumpet Honeysuckle

Lonicera sempervirens

Trumpet honeysuckle is a twining, climbing vine that is easily trained on a trellis, arbor, or through a garden shrub. It is deciduous in mild winters except it will hold on to its leaves. Its early spring flowers, although not fragrant, are a beautiful combination of red or coral on the outside with yellow-orange inside. Red fruits follow in summer. The leaves turn a dark, bluish green as they mature.

J F M A M J J A S O N D

Propagation is by cuttings or seed, and trumpet honeysuckle prefers a moist, well-drained soil, although it will tolerate some drought.

Trumpet honeysuckle's colorful flowers attract swallowtail butterflies and ruby-throated hummingbirds, while northern cardinals and gray catbirds feed on the fruits.

summer fruit

FIELD NOTES

■ Zones 6–9

■ Deciduous

■ 10–20' (3–6 m) or higher

☀ Full sun–part shade

◊ Moderate

tiger swallowtail butterfly

Jackson Vine

Smilax smallii

Jackson vine is a tough vine with handsome, glaucous foliage and, when they are ripe in fall, dark, almost black, berries. A fast grower, it is also easy to control. It is a choice vine for an evergreen screen on a trellis or fence, or as an accent in the garden. Jackson vine grows in various habitats, including sandy woodlands. Leaves are used in the florist trade because they hold their color and shape long after cutting.

Propagation is by division and cuttings, and Jackson vine prefers a rich, moist, well-drained soil, although it will tolerate some drought.

J F M A M J J A S O N D

A number of birds are attracted to the fruit of this vine, including gray catbirds, mockingbirds, robins, flickers, brown thrashers, thrushes, pileated woodpeckers, northern cardinals, sparrows, cedar waxwings, grouse, turkeys, and waterfowl. Lizards, such as the agile green anole, can be found climbing among the vines.

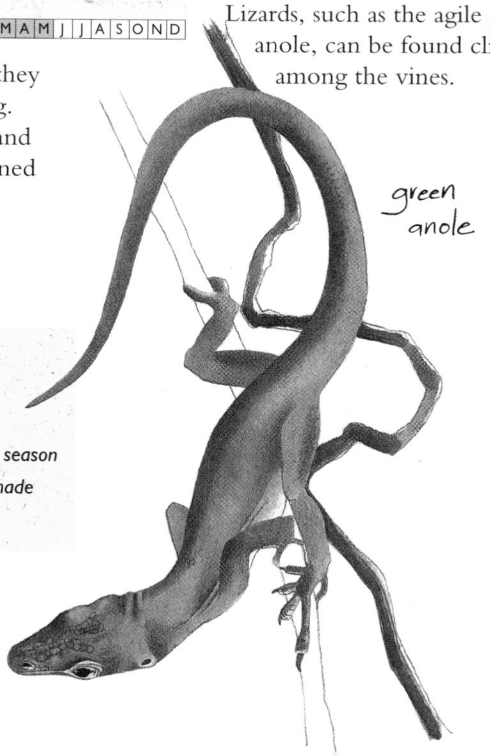

green anole

FIELD NOTES
- Zones 7–9
- Evergreen
- 20–30' (6–9 m) in a season
- ☀ Full sun–part shade
- ◌ Moderate

271

Daylily (left); sourwood (right)

Additional species

Plant name/flower color	Zone	Height	Visitors/attraction/season
HERBACEOUS FLOWERS			
Swamp sunflower (yellow) *Helianthus angustifolius*	6–9	6–8' (1.8–2.4 m)	butterflies/flowers/summer–fall birds/seeds/late summer–fall
Daylily (many) *Hemerocallis* spp. (exotic)	3–10	1–3' (30–90 cm)	butterflies/flowers/summer
Dame's rocket (white, purple) *Hesperis matronalis* (exotic)	3–9	3–4' (90–120 cm)	butterflies/flowers/spring, summer
Virginia bluebells (blue/pink) *Mertensia virginica*	3–10	2' (60 cm)	hummingbirds/flowers/spring
Flowering tobacco (white) *Nicotiana alata* (exotic)	all (annual)	2–4' (60–120 cm)	moths/flowers/summer–fall
Showy stonecrop (pink, red, white) *Sedum spectabile* (exotic)	3–10	24' (7.3 m)	butterflies/flowers/summer
SHRUBS			
Glossy abelia (white/pink) *Abelia* x *grandiflora* (exotic)	5–10	6' (1.8 m)	butterflies/flowers/summer, fall
Butterfly bush (purple, yellow, white) *Buddleia davidii* (exotic)	5–10	10' (3 m)	butterflies/flowers/summer, fall
Summersweet clethra (white) *Clethra alnifolia*	3–9	3–8' (90–240 cm)	butterflies/flowers/summer
Climbing hydrangea (white) *Decumaria barbara*	6–9	30' (9 m)	squirrels/shelter/spring
Heliotrope (violet/purple) *Heliotropium arborescens* (exotic)	9–10	6–8' (1.8–2.4 m)	butterflies/flowers/summer
Black haw (white) *Viburnum prunifolium*	3–9	15' (4.5 m)	birds/fruits/fall
TREES			
Devil's walking stick (white) *Aralia spinosa*	4–9	20' (6 m)	birds/fruits/fall
Common persimmon (white) *Diospyros virginiana*	4–9	35' (10.5 m)	oppossum, deer/fruits/fall
Sweetgum (green) *Liquidambar styraciflua*	5–10	60–100' (18–30 m)	birds/seeds/fall–winter birds/nesting/spring–summer
Sourwood (white) *Oxydendrum arboreum*	5–9	30' (9 m)	bees/flowers/summer
Longleaf pine (n/a) *Pinus palustris*	5–10	100–130' (30–39 m)	birds, mammals/seeds/fall birds/nesting/spring–summer
Chaste tree (white, purple, pink, blue) *Vitex agnus-castus* (exotic)	7–10	8–10' (2.4–3 m)	butterflies/flowers/summer–fall
VINES			
Moonflower (white) *Ipomoea alba* (exotic)	9–10	10' ± (3 m) ±	moths/flowers/summer
Passion flower (lavender/white) *Passiflora incarnata*	5–10	20' (6 m)	butterflies/flowers/summer caterpillars/leaves/summer

RESOURCES
DIRECTORY

FURTHER READING

The books listed below are divided into those containing general gardening information and those with information relating to the six regions of The Backyard Habitat. Some field guides, however, cover more than one of the regions, so be sure to check the entries for neighboring regions as well as your particular region.

General Gardening

The Audubon Society Nature Guides: including *Grasslands; Wetlands; Western Forests; Eastern Forests.* Various authors (Alfred A. Knopf).

Collecting, Processing, and Germinating Seeds of Wildland Plants, by James A. and Cheryl Young (Timber Press, 1986).

Commonsense Pest Control, by William and Helga Olkowski, and Sheila Daar (Taunton Press, 1991).

Earthly Pleasures, by Roger Swain (Scribners, 1981).

The Encyclopedia of Organic Gardening, by Rodale Press (Rodale Press, 1978). One of the best and most complete books on organic gardening.

Gardening by Mail: a Source Book, by Barbara Barton, (Houghton Mifflin, 1994).

Gardening Success with Difficult Soils, by Scott Ogden (Taylor Publishing Co., 1992).

Gardening with Wildflowers and Native Plants, edited by Claire E. Sawyers (Brooklyn Botanic Garden, 1989).

Going Native: Biodiversity in our own Backyards, edited by Janet Marinelli. Handbook #140 (Brooklyn Botanical Garden, 1994).

Groundwork, by Roger Swain (Houghton Mifflin, 1994).

Growing and Propagating Showy Native Woody Plants, by Richard E. Bir (University of North Carolina Press, 1992).

Growing and Propagating Wildflowers, by Harry R. Phillips (University of North Carolina Press, 1985).

Illustrated Encyclopedia of Perennials, by Ellen Phillips and C. Colston Burrell (Rodale Press, 1993).

Landscaping with Wildflowers, by Jim Wilson (Houghton Mifflin, 1992).

The National Wildlife Research Center's Wildflower Handbook, by the N.W.R.C. (Voyageur Press, 1992).

Native Trees, Shrubs, and Vines for Urban and Rural America, by Gary L. Hightshoe (Van Nostrand Reinhold, 1988).

The Natural Habitat Garden, by Ken Druse (Clarkson N. Potter, 1994).

Natural Landscaping, by John Diekelmann and Robert Schuster (McGraw Hill, 1982).

Pests of Landscape Trees and Shrubs: An Integrated Pest Management Guide, by Steve Dreistadt (University of California Press, 1994).

Please Don't Eat My Garden! by Nancy McCloud (Sterling, 1992).

Propagation of Wildflowers, by Will C. Curtis, revised by William E. Brumback (New England Wild Flower Society, 1986).

Rodale's Garden Insect, Disease, and Weed Identification Guide, by Miranda Smith and Anna Carr (Rodale Press, 1988).

Taylor's Guide to Natural Gardening, edited by Roger Holmes (Houghton Mifflin, 1993).

Taylor's Guide to Specialty Nurseries, by Barbara Barton (Houghton Mifflin, 1993).

Taylor's Master Guide to Gardening, edited by Rita Buchanan and Roger Holmes (Houghton Mifflin, 1994).

The Trees of North America, by Alan Mitchell (Facts on File, 1987).

Wildflower Gardener's Guide: Northeast, Mid-Atlantic, Great Lakes and Eastern Canada; California, Desert Southwest, and Northern Mexico; Midwest, Great Plains, and Canadian Prairies. (3 volumes), by Henry W. Art (Storey Communications, 1987, 1990, 1991).

Wildflowers in Your Garden: A Gardener's Guide, by Vicki Ferreniea (Random House, 1993).

Wildlife Gardening

Animal Architecture, by Karl von Frisch (Harcourt Brace Jovanovich, 1974).

Attracting Backyard Wildlife, by William J. Merilees (Voyageur Press, 1989).

Attracting Birds to your Backyard, by C. Heimerdinger and S.H. Spofford (Publications Intel Ltd., 1991). Recommended by the American Birding Association.

The Audubon Society Field Guide to Mammals, by John O. Whitaker (Alfred A. Knopf, 1980).

The Audubon Society Field Guide to North American Butterflies, by Robert Mitchell Pyle (Alfred A. Knopf, 1992).

The Backyard Naturalist, by Craig E. Tufts (The National Wildlife Federation, 1988).

Backyard Wildlife Habitat Information Packet, by the National Wildlife Federation.

Broadsides from Other Orders: A Book of Bugs, by Sue Hubbell (Random House, 1993).

The Butterfly Garden, by Jerry Sedenko (Villard Books, 1991).

Butterfly Gardening: Creating Summer Magic in Your Garden, by the Xerxes Society/ Smithsonian Institution (Sierra Club Books, 1990).

The Evening Garden, by Peter Loewer (Macmillan, 1993).

The Field Guide to Wildlife Habitats of the Eastern United States, by Janine Benyus (Fireside Books, 1989).

A Field Guide to Your Own Backyard, by John Hanson Mitchell (W. W. Norton and Co., 1985). Seasonally divided guide to plants and animals commonly found around suburban and rural homes.

How to Attract Hummingbirds and Butterflies, by John Dennis and Mathew Tekulsky (Ortho Books, 1991).

Landscaping for Wildlife, by Carrol L. Henderson (Minnesota Dept. of Natural Resources, 1987).

The Naturalist's Garden, by Ruth Shaw Ernst (The Globe Pequot Press, 1993).

The Naturalist's Path, Beginning the Study of Nature, by Cathy Johnson (Walker Books, 1991).

Observing Insect Lives, by Donald and Lillian Stokes (Little Brown, 1983).

Songbirds in Your Garden, by John K. Terres (Algonquin Books of Chapel Hill, 1994).

Suburban Nature Guide: How to Discover and Identify the Wildlife in Your Backyard, by David Mohrhardt and Richard E. Schinkel (Stackpole Books, 1991).

The Wildlife Gardener, by John Dennis (Alfred A. Knopf, 1985).

Wildlife in Your Garden, by Gene Logsdon (Rodale Press, 1974).

Your Backyard Wildlife Garden, by Marcus Shneck (Rodale Press, 1992).

The Backyard Habitat
THE WEST COAST

Butterflies of the San Francisco Bay Region, by J.W. Tilden (University of California Press, 1965). Comprehensive coverage; some color photographs.

California Insects, by Jerry Powell and Charles L. Hogue (University of California Press, 1979). Classification and details of common species.

Complete Garden Guide to Native Perennials of California, by Glenn Keator (Chronicle Books, 1992). Information on growing a wide range of native perennials; line drawings.

Complete Garden Guide to Native Shrubs of California, by Glenn Keator (Chronicle Books, 1994).

Gardener's Guide to California Wildflowers, by Kevin Connelly (Theodore Payne Foundation, 1991).

Growing California Native Plants, by Marjorie G. Schmidt. (University of California Press, 1980).

Landscape Plants for Western Regions, by Robert C. Perry (Land Design Publishing, 1992).

Sunset Western Garden Book, by Sunset Books (Lane Publishing Co., 1988). A very helpful book with an especially useful section on landscape plants.

MOUNTAINS AND BASINS

From Grassland to Glacier: The Natural History of Colorado, by Cornelia Fleischer and John C. Emerick (Johnson Books, 1984).

Medicinal Plants of the Mountain West, by Michael Moore (Museum of New Mexico Press, 1989).

Native Plants of Genesee and How to Use Them in Foothills Residential Landscape Design, by Sylvia B. Brockner and Jeanne R. Janish (Genesee Foundation, 1987).

Rocky Mountain Wild Flowers, by A. E. Porsild (National Museums of Canada, 1979).

Shrubs and Trees of the Southwest Uplands, by Francis H. Elmore (Southwest Parks and Monuments Association, 1976). Good color photographs, useful drawings, and informative text.

Wildflowers, Grasses, and other Plants of the Northern Great Plains, by Theodore Van Bruggen (Badlands Natural History Association, 1983). Many good photographs.

The Xeriscape Flower Garden: A Waterwise Guide for the Rocky Mountains Region, by Jim Knopf (Johnson Books, 1991). Useful for all western desert regions.

THE DESERT SOUTHWEST

Arizona Flora, by Thomas H. Kearney and Robert H. Peebles (University of California Press, 1960).

Desert Plants: Biotic Communities of the American Southwest—United States and Mexico, edited by David E. Brown (The University of Arizona, 1982), Vol. 4, Nos. 1–4. Habitat restoration.

How to Grow Native Plants of Texas and the Southwest, by Jill Nokes (Texas Monthly Press, 1986). Detailed propagation and cultivation information; watercolor illustrations.

Native Gardens for Dry Climates, by Sally and Andy Wasowski. (Clarkson N. Potter, 1995). Garden plans, indigenous charts, and plant profiles.

Southwestern Landscaping with Native Plants, by Judith Philips (Museum of New Mexico Press, 1987).

Taylor's Guide to Gardening in the Southwest, edited by Roger Holmes and Rita Buchanan (Houghton Mifflin, 1992). Includes the West Coast.

Trees and Shrubs of the Southwestern Deserts, by Lyman Benson and Robert A. Darrow (University of Arizona Press, 1981).

Trees, Shrubs, and Woody Vines of the Southwest, by Robert A. Vines (University of Texas Press, 1960).

Woody Plants of the Southwest, by Samuel H. Lamb (Sunstone Press, 1989). Includes maps and black and white photographs.

THE PRAIRIES

Flora of the Great Plains, by the Great Plains Flora Association (University Press of Kansas, 1986). A comprehensive botanical reference.

Jewels of the Plains: Wildflowers of the Great Plains Grasslands and Hills, by Claude A. Barr (University of Minnesota Press, 1983).

Native Texas Plants: Landscaping Region by Region, by Sally and Andy Wasowski (Texas Monthly Press, 1988). Includes the Desert Southwest.

The Prairie Garden, by Robert I., and Beatrice S. Smith (University of Wisconsin Press, 1980).

Prairie Wildflowers, by R. Currah, A Smreciu, and M. Van Dyk. (University of Alberta, 1983).

Roadside Wildflowers of the Southern Great Plains, by Craig C. Freeman and Eileen K. Schofield (University Press of Kansas, 1991).

The Shortgrass Prairie, by Ruth Carol Cushman and Stephen R. Jones (Pruett Publishing Company, 1989).

Wildflowers of Nebraska and The Great Plains, by Jon Farrar (Nebraskaland Magazine, Nebraska Game and Fish Commission, 1990). Contains many color photographs.

Wildflowers of the Tallgrass Prairie, by Sylvan T. Runket and Dean M. Roosa (Iowa State University Press, 1989).

THE NORTHEAST

A Field Guide to Wildflowers of Northeastern and North-Central North America, by Roger Tory Peterson and Margaret McKenny (Houghton Mifflin, 1968). Includes the Southeast and Prairies. Comprehensive coverage of the regions' wildflowers.

Landscape Plants for Eastern North America, by H. L. Flint (John Wiley, 1983).

Meadows and Meadow Gardening, by New England Wild Flower Society (N.E.F.S., 1990).

Native Plants for Woodland Gardens: Selection, Design and Culture, by New England Wild Flower Society (N.E.F.S., 1987). Includes the Southeast.

The Wildflower Meadow Book: A Gardener's Guide, by Laura C. Martin (Globe Pequot Press, 1990).

THE SOUTHEAST

Butterfly Gardening for the South, by Geyata Ajilvsgi (Taylor Publishing Company, 1990). Includes the Southwest region. Comprehensive text with great photographs.

Gardening with Native Plants of the South, by Sally and Andy Wasowski.(Taylor Publishing Company, 1994). Includes plants of the Desert Southwest.

Landscape Plants of the Southeast, by R. G. Halfacre and A. R. Shawcroft (Sparks Press, 1989).

Native Shrubs and Woody Vines of the Southeast: Landscaping Uses and Identification, by Leonard E. Foote and Samuel B. Jones, Jr. (Timber Press, 1989). Good photographs and some information on wildlife species.

A Natural History of Trees: Eastern and Central North America, by Donald Culross Peattie (Houghton Mifflin, 1991). Good history and natural history content.

Taylor's Guide to Gardening in the South, edited by Rita Buchanan and Roger Holmes (Houghton Mifflin, 1992).

Trees and Shrubs for the Southeast, by B. Wiggington (University of Georgia Press, 1963).

Wildflowers of the Southeastern United States, by Wilbur H. Duncan and Leonard E. Foote (University of Georgia Press, 1975).

PLANT SOURCES *and* ORGANIZATIONS

This list includes nurseries and seed companies, and wildlife and conservation services and associations. The following abbreviations have been used for the nurseries and seed companies: MO (mail order), NUR (retail nursery), GDN (demonstration garden, open to public), and SASE (self-addressed stamped envelope).

Nurseries and Seed Companies

WESTERN UNITED STATES

Arizona

Homan Brothers Seed (Seeds), P.O. Box 337, Glendale, AZ 85311. ph. (602) 244-1650 Catalog: 2 first class stamps MO, NUR, GDN (by appointment). Habitat collected seeds of desert plants

California

Clyde Robin Seed Co. (Seeds), 3670 Enterprise Ave. Hayward, CA 94545. ph. (510) 785-0425 MO. Wildflower mixes for many regions

Larner Seeds (Plants and seeds), PO Box 407, 235 Grove Rd. Bolinas, CA 94924. ph. (415) 868-9407 Catalog: $2 MO, NUR, GDN (Tue and Fri). Trees, shrubs, vines, grasses, and wildflowers

Las Pilitas Nursery (Plants and seeds), Star Route, Box 23X, Las Pilitas Rd. Santa Margarita, CA 93453. ph. (805) 438-5992 Catalog: $6 (free price list) MO, NUR, & GDN (Sat except holidays). Very broad selection of California native plants

Mockingbird Nursery (Plants), 1670 Jackson St. Riverside, CA 92504. ph. (909) 780-3571 Plant list free. NUR (by appointment only). Trees and shrubs

Moon Mountain Wildflowers (Seeds), PO Box 725, Carpinteria, CA 93014 Catalog: $3 MO. Wildflowers.

Mostly Natives Nursery (Plants), 27235 Highway 1, Tomales, CA 94971. ph. (707) 878-2009 Catalog: $3 MO, NUR, GDN (Wed–Sun). Trees, shrubs, grasses, and wildflowers

The Theodore Payne Foundation (Plants and seeds), 10459 Tuxford St. Sun Valley, CA 91352. ph. (818) 768-1802 Catalog: $2 MO (Seeds only), NUR, GDN (Wed–Sun). Trees, shrubs, grasses, and wildflowers

Van Ness Water Gardens (Plants), 2460 N. Euclid Ave. Upland, CA 91784. ph. (909) 982-2425 Catalog: free. MO, NUR, GDN (Tue–Sat). Water and bog plants

Western Hills Nursery (Plants), 16250 Coleman Valley Rd. Occidental, CA 95465. ph. (707) 874-3731 NUR, GDN (Feb–Nov, Thur–Sun). Native plants, berrying and butterfly plants

Yerba Buena Nursery (Plants and seeds), 19500 Skyline Boulevard, Woodside, CA 94062. ph. (415) 851-1668 Catalog: $1 NUR (daily except holidays). Trees, shrubs, ferns, grasses, and wildflowers

Colorado

Neils Lunceford (Plants and seeds), PO Box 2130, 740 Blue River Parkway, Silverthorne, CO 80498. ph. (303) 468-0340 Catalog: free. MO (Seed only), NUR, GDN (Mon–Fri, summer daily). Trees, shrubs, wildflowers

Montana

Valley Nursery (Plants), PO Box 4845, 2801 N. Montana Ave. Helena, MT 59604. Catalog: 2 first class stamps MO, NUR, GDN (daily in season or by appointment). Hardy trees and shrubs

New Mexico

A High Country Garden (Plants), 2902 Rufina St. Santa Fe, NM 87501. ph. (505) 438-3031 Catalog: free MO, NUR, GDN (daily). Native shrubs, grasses, and wildflowers.

Agua Fria Nursery (Plants), 1409 Agua Fria St. Santa Fe, NM 87501. ph. (505) 983-4831 Catalog: free. MO, NUR (daily). Wildflowers

Bernardo Beach Native Plant Farm (Plants and seeds), 520 Montano Road, N.W. Albuquerque, NM 87107. ph. (505) 345-6248 NUR (Mon–Sat). Broad selection of native plants

Desert Moon Nursery (Plants), PO Box 600, Veguita, NM 87062. Catalog: $1 MO, NUR, GDN (by appointment). Cactus and succulent plants, wildflowers

Plants of the Southwest (Plants and seeds), Route 6, Box 11A, Agua Fria, Santa Fe, NM 87501. ph. (505) 471-2212 Catalog: $3.50 (price list free) MO, NUR (daily). Trees, shrubs, grasses, and wildflowers

Oregon

Callahan Seeds (Seeds), 6045 Foley Lane, Central Point, OR 97502. ph. (503) 855-1164 MO, NURS (by appointment). Broad selection of Northwestern trees and shrubs

Forestfarm (Plants), 990 Tetherow Road, Williams, OR 97544. ph. (503) 846-6963 Catalog: $3 MO, NUR (by appointment). Very broad selection of trees, shrubs, and wildflowers.

Russell Graham (Plants), 4030 Eagle Crest Rd. NW Salem, OR 97304. ph. (503) 362-1135 Catalog: $2 MO, NUR, GDN (Sat. by appointment). North-western ferns and wildflowers

Siskiyou Rare Plant Nursery (Plants), 2825 Cummings Rd. Medford, OR 97501. ph. (503) 772-6846 Catalog: $2 MO, NUR, GDN (first and last Saturday Mar–Nov, or by appointment). Small shrubs, rock garden, and alpine plants

Washington

Colvos Creek Nursery (Plants), PO Box 1512, Vashon Island, WA 98070. ph. (206) 441-1509 Catalog: $2 MO, NUR (by appointment). Northwestern native trees, shrubs, and wildflowers

Frosty Hollow (Seeds), PO Box 53, Langley, WA 98260. ph. (206) 579-2332 MO/Long SASE. Northwestern trees, shrubs, grasses, and wildflowers

Plants of the Wild (Plants), PO Box 866, Willard Field, Tekoa, WA 99033. ph. (509) 284-2848 Catalog: free. MO, NUR, GDN (by appointment). Trees, shrubs, grasses, and wildflowers

CENTRAL UNITED STATES

Arkansas

Holland Wildflower Farm (Seeds), 290 O'Neal Lane, Elkins, AR 72727. ph. (501) 643-2622 Seed list: free, Growing guide: $2.50. MO, NUR, GDN (by appointment). Prairie wildflowers

Illinois

Bluestem Prairie Nursery (Plants and seeds), Route 2, Box 106A,

Hillsboro, IL 62049.
  ph. (217) 532-6344
  Catalog: free MO, GDN (by
  appointment). Prairie grasses and
  wildflowers
Iowa
Ion Exchange (Plants and Seeds),
  1878 Old Mission Drive,
  Harpers Ferry, IA 52146.
  ph. (319) 535-7231. Catalog: free
  MO. Grasses and wildflowers
Kansas
Sharp Bros. Seed Co. (Seeds),
  PO Box 140, Healy, KS 67850.
  ph. (316) 398-2231
  Catalog: $5. MO. Midwestern
  grasses and wildflowers
Louisiana
Gulf Coast Plantsmen (Plants),
  15680 Perkins Road, Baton
  Rouge, LA 70810.
  ph. (504) 751-0395
  NUR (Tue–Sat, Sun am).
  Trees, shrubs, wildflowers
Louisiana Nursery (Plants),
  Route 7, Box 43, Opelousas,
  LA 70570. ph. (318) 948-3696
  Catalogs: $5 each (specify plant
  type). MO, NUR, & GDN
  (by appointment). Very broad
  selection of Southern native plants
Minnesota
Prairie Moon Nursery (Plants and
  seeds), Route 3, Box 163,
  Winona, MN 55987
  ph. (507) 452-1362. Catalog: $2
  MO, NUR, GDN (By
  appointment). Prairie grasses and
  wildflowers
Missouri
Missouri Wildflowers Nursery (Plants
  and seeds), 9814 Pleasant Hill Rd.
  Jefferson City, MO 65109.
  ph. (314) 496-3492. Catalog: $1
  MO, NUR (spring, daily; fall,
  weekends). Grasses and wildflowers
Nebraska
Stock Seed Farms, Inc. (Seeds),
  28008 Mill Rd., Murdock,
  NE 68407. ph. (402) 867-3771
  Catalog: free. MO.
  Prairie grasses and wildflowers
Texas
Lilyponds Water Gardens (Plants),
  839 FM 1489, Brookshire,
  TX 77423. ph. (713) 934-8525
  Catalog: free (see Maryland
  address). NUR, GDN (daily,
  Mon–Sat in winter).
  Water and bog plants
Turner Seed Co. (Seeds),
  Route 1, Box 292, Breckenridge,
  TX 76424. ph. (817) 559-2065
  Catalog: free. MO. Grasses and
  wildflowers (seed in bulk).
Wildseed, Inc. (Seeds), PO Box 308,
  Eagle Lake, TX 77434.
  ph. (409) 234-7353. Catalog: free
  MO. Wildflowers for all regions
Yucca Do Nursery (Plants),
  PO Box 655, Waller, TX 77484.
  ph. (409) 826-6363. Catalog: $3
  MO. Native plants of the
  Southwest and Northern Mexico

Wisconsin
Country Wetlands Nursery and
  Consulting (Plants and seeds),
  S. 75 W. 20755, Field Drive,
  Muskego, WI 53150.
  ph. (414) 679-1268. Catalog: $2
  MO, NUR, & GDN (Mon–Fri,
  call ahead). Bog and wetland plants
Kester's Wild Game Food Nurseries,
  Inc. (Plants and seeds),
  PO Box 516, Omro, WI 54963.
  ph. (414) 685-2929
  Catalog: $3. MO, NUR
  (Mon–Fri, call ahead). Water and
  wetland plants, grasses
Little Valley Farm (Plants),
  5693 Snead Creek Rd. Spring
  Green, WI 53599.
  ph. (608) 935-3324
  Catalog: 1 first class stamp
  MO, NUR (by appointment).
  Trees, shrubs, vines, wildflowers
Prairie Nursery (Plants and seeds),
  PO Box 306, 3291 Dyke Ave.
  Westfield, WI 53964.
  ph. (608) 296-3679. Catalog: $3
  MO, GDN (by appointment)
  Grasses and wildflowers
Wildlife Nurseries, Inc. (Plants and
  seeds), PO Box 2724, Oshkosh,
  WI 54903. ph. (414) 231-3780
  Catalog: $2. MO, NUR (Mon–Fri
  by appointment). Wetland plants,
  grasses, and wildflowers

EASTERN UNITED STATES
Florida
NWN Nursery (Plants and seeds),
  1365 Watford Circle, Chipley,
  FL 32428. ph. (904) 638-7572
  Catalog: free. MO, NUR
  (Tue–Fri or by appointment).
  Trees, shrubs, and vines
Native Nurseries (Plants and seeds),
  1661 Centerville Rd. Tallahassee,
  FL 32308. ph. (904) 386-8882.
  Plant list free. NUR, GDN
  (Mon–Sat). Trees, shrubs, and
  wildflowers
Georgia
Goodness Grows (Plants),
  PO Box 311, Highway 77 North
  Lexington, GA 30648

ph. (706) 743-5055. Catalog: free
  MO, NUR, GDN (Mon-Sat, Sun
  in spring). Wildflowers
Transplant Nursery (Plants),
  Route 2, Parkertown Rd.
  Lavonia, GA 30553.
  ph. (706) 356-8947
  Catalog: free. NUR, GDN
  (Mon–Sat). Shrubs
Kentucky
Shooting Star Nursery (Plants and
  seeds), 444 Bates Road, Frankfort,
  KY 40601. ph. (502) 223-1679
  Catalog $2 MO, NUR (May-Oct:
  Mon–Sat, call ahead). Trees,
  shrubs, ferns, and wildflowers
Maine
Conley's Garden Center (Plants),
  145 Townsend Ave. Boothbay
  Harbor, ME 04538.
  ph. (207) 633-5020.
  NUR, GDN (daily, Mon–Sat
  in Jan–Mar). Native trees, shrubs,
  and groundcovers
Maryland
Kurt Bluemel (Plants), 2740 Greene
  Lane, Baldwin, MD 21013.
  ph. (410) 557-7229
  Catalog: $3. MO. Grasses,
  bamboos, ferns, and wildflowers
Lilypons Water Gardens (Plants),
  6800 Lilypons Rd. PO Box 10,
  Buckeystown, MD 21717.
  ph. (310) 874-5133
  Catalog: free. MO, NUR, GDN
  (daily, Mon–Sat in winter).
  Water and bog plants
Massachusetts
Triple Brook Farm (Plants),
  37 Middle Rd. Southampton,
  MA 01073.
  ph. (413) 527-4626 (evenings)
  MO, NUR & GDN (by appoint-
  ment). Shrubs and wildflowers
Michigan
Michigan Wildflower Farm (Seeds),
  11770 Cutler Rd. Portland,
  MI 48875. ph. (517) 647-6010
  Catalog: free. MO. Native
  Michigan wildflowers
Oikos Tree Crops (Plants),
  PO Box 19425, Kalamazoo,
  MI 49019. ph. (616) 624-6233
  Catalog: $1. MO, NUR, and
  GDN (by appointment). Hardy
  nut trees, oaks, and shrubs
Wavecrest Nursery & Landscaping
  Co. (Plants), 2509 Lakeshore Dr.
  Fennville, MI 49408.
  ph. (616) 543-4175
  MO, NUR, GDN (daily Mar–
  Nov). Native trees and shrubs
New Hampshire
Oakridge Nursery (Plants), PO Box
  182, East Kingston, NH 03827.
  ph. (603) 642-8227.
  Catalog: free MO, NUR
  (call ahead). Trees, shrubs,
  ferns, wildflowers
New Jersey
Lofts Seed (Seed), PO Box 146
  Bound Brook, NJ 08805.
  ph. (800) 526-3890. Catalog: free
  MO. Grasses and wildflowers

Wild Earth Native Plant Nursery
(Plants), 49 Mead Ave. Freehold,
NJ 07728. ph. (908) 308-9777
Catalog: $2. MO, NUR
(call ahead; not at above address).
Grasses, ferns, and wildflowers

North Carolina
Gardens of the Blue Ridge,
PO Box 10, 9056 Pittman Gap
Rd. Pineola, NC 28662
ph. (704) 733-2417. Catalog: $3
MO, NUR, GDN (Mon–Sat am,
call ahead). Trees, shrubs, ferns,
and wildflowers
Holbrook Farm & Nursery (Plants),
PO Box 368, 115 Lance Rd.
Fletcher, NC 28732.
ph. (704) 891-7790. Catalog: free
MO, NUR, GDN (Mon–Sat).
Southeastern wildflowers
Niche Gardens (Plants),
1111 Dawson Road,
Chapel Hill, NC 27516.
ph. (919) 967-0078
Catalog: $3. MO, NUR, GDN
(Tue–Fri, some spring weekends).
Southeastern native plants and
wildflowers

Pennsylvania
Appalachian Wildflower Nursery
(Plants), Route 1, Box 275A,
Reedsville, PA 17084.
Catalog: $2. MO, NUR (write to
make an appointment). Mid-
Atlantic native plants
Jacob's Ladder Natural Gardens
(Plants), 1345 Conshohocken State
Road, PO Box 145, Gladwyn, PA
19035. ph. (215) 525-6773
NUR (Mon–Fri). Trees, shrubs,
and wildflowers
Musser Forests (Plants),
PO Box 340, Route 119 North
Indiana, PA 15701.
ph. (412) 465-5686
Catalog: free. MO, NUR (daily).
Trees and shrubs

South Carolina
Woodlanders, Inc. (Plants),
1128 Colleton Avenue,
Aiken, SC 29801.
Catalog: $2 MO, NUR, GDN (by
appointment). Broad selection of
Southeastern native plants

Tennessee
Native Gardens (Plants and seeds),
5737 Fisher Lane, Greenback,
TN 37742. ph. (615) 856-0220
Catalog: $2
MO, NUR, GDN (by
appointment). Trees, shrubs, and
wildflowers
Sunlight Gardens (Plants),
174 Golden Lane,
Andersonville, TN 37705.
ph. (615) 494-8237
Catalog: $3. MO, NUR (by
appointment) Southeastern and
Northeastern wildflowers

Vermont
Putney Nursery (Seeds),
P O Box 265, Putney, VT 05346.
ph. (802) 387-5577
MO, NUR (Mon–Sat, Apr–Dec).

Wildflower seeds, plants at nursery
Vermont Wildflower Farm (Seeds),
PO Box 5, Route 7
Charlotte, VT 05445.
ph. (802) 425-3500
Catalog: free. MO, NUR, GDN
(May–Oct daily). Wildflower seeds
for all regions

CANADA
Aimers (Seeds), 81 Temperence
Street, Aurora, ON L4G 2R1.
ph. (905) 833-5282. Catalog: $4.
MO, Shop, (daily). Wildflower
seeds.
Fraser's Thimble Farm (Plants),
175 Arbutus Road, Saltspring
Island, BC V8K 1A3.
ph. (604) 537-5788.
MO, NUR (daily). Native trees,
shrubs, and wildflowers of the
Pacific Northwest.
Keith Somers Trees (Plants and
seeds), 10 Tillson Ave.
Tillsonburg, ON N4G 2Z6.
ph. (519) 842-5148.
Catalog $2. MO, NUR (May-
Oct, Mon–Sat or call ahead).
Native trees and shrubs,
wildflower plants, and seeds
Windmill Point Farm and Nursery
(Plants), 2103 Boul. Perrot, ND
Ile Perrot, Quebec J7V 8P4. ph.
(514) 453-9757. MO, NUR
(April-Oct, Sat-Sun, call ahead).
Native trees, shrubs, and herbs

Wildlife and
Conservation Societies
UNITED STATES
American Birding Association
PO Box 6599, Colorado Springs,
CO 80934. ph. (719) 634-7736
Birding (bi-monthly)
American Nature Study Society,
5881 Cold Brook Road, Homer,
NY 13077. ph. (607) 749-3655
Nature Study (irregularly)
ANSS News (quarterly)
Bat Conservation International, PO
Box 162603, Austin, TX 78716.
ph. (512) 327-9721
Bats (quarterly)
Lepidopterists' Society,
c/o Julian Donohue, Natural
History Museum, 900 Exposition
Boulevard, Los Angeles,
CA 30007. ph. (213) 744-3364
Journal (quarterly),
Newsletter (bi-monthly)
National Audubon Society,
950 Third Avenue, New York,
NY 10022. ph. (212) 832-3200
Audubon Magazine, American Birds
(both bi-monthly)
National Institute of Urban Wildlife,
PO Box 3015, Shepherdstown,
WV 25443. ph. (304) 274-0205
Urban Wildlife News (quarterly)
National Wildlife Research
Center, 2600 FM 973, North
Austin, TX 78725.
ph. (512) 929-3600
Wildflower (bi-monthly)

National Wildlife Federation
1400 16th Street NW,
Washington, DC 20036
ph. (202) 797-6800
National Wildlife (bi-monthly)
North American Bluebird Society,
PO Box 6295,
Silver Spring, MD 20906
ph. (301) 384-2798
Sialia (quarterly)
North American Butterfly
Association, 4 Delaware Road,
Morristown, NJ 07960
American Butterflies (quarterly), *The
Anglewing* (biannually)
Purple Martin Conservation
Association,
c/o James R. Hill III, Edinboro
Univ. of Pennsylvania, Edinboro,
PA 16444. ph. (814) 734-4420
Purple Martin Update (quarterly)
Wildlife Forever,
PO Box 3404, 12301 Whitewater
Drive, Suite 210, Minnetonka,
MN 55343. ph. (612) 936-0605
Cry of the Wild! (quarterly)
Xerces Society (Butterflies),
c/o Melody Allen, 10 SW Ash
Street, Portland, OR 97204
ph. (503) 222-2788
Wings (twice a year)

CANADA
Canadian Nature Federation,
1 Nicholas Street, Suite 520,
Ottawa, ON, Canada K1N 7B7.
ph. (613) 562-2447
Nature Canada (quarterly)
Nature Alert (quarterly)
The Canadian Wildflower Society,
Unit 12A, Box 228, 4981
Highway #7 East, Markham, ON,
L3R 1N1. ph. (905) 294-9075
Wildflower (quarterly)
Canadian Wildlife Federation, 2740
Queenview Drive, Ottawa, ON,
K2B 1A2. ph. (613) 721-2286
International Wildlife (bi-monthly)
English and French editions.
Ranger Rick's Nature Magazine
(monthly)
Big Backyard (monthly) Ages 3–6

NOTE: Most states have
Native Plant Societies.
Your local library could
help you find the one for
your state, or check in
*Gardening by Mail:
a Source Book* (see
Further Reading).

PLACES *of* INSPIRATION

The following listing provides details of inspirational botanical gardens throughout North America and Canada. Public gardens, particularly those featuring native plants, can be a wonderful source of inspiration when planning your natural garden.

WESTERN UNITED STATES

Anza-Borrego Desert State Park,
    PO Box 299, Borrego Springs, CA
    92004. ph. (619) 767-4684
Arizona-Sonora Desert Museum,
    2021 N. Kinney Rd. Tucson, AZ
    85743. ph. (602) 883-2702
The Bloedel Reserve, 7571 NE
    Dolphin Dr. Bambridge Island,
    WA 98110. ph. (206) 842-7631
Bryce Thompson Southwestern
    Arboretum, PO Box AB,
    Superior, AZ 85273.
    ph. (602) 941-1217
Denver Botanic Garden,
    909 York St, Denver,
    CO 80206-3799.
    ph. (303) 331-4000
Leach Botanical Garden,
    6704 SE 122nd Ave, Portland,
    OR 97236. ph. (503) 761-9503
The Living Desert, 47900 Portola
    Ave, Palm Desert, CA 92260.
    ph. (619) 246-5694
Moorten Botanical Garden,
    1702 South Palm Canyon Dr.
    Palm Springs, CA 92264.
    ph. (619) 327-6555
New Mexico State University
    Botanical Garden, Dept. of
    Agronomy and Horticulture,
    Box 3Q, Las Cruces,
    NM 88003. ph. (505) 649-3638
Red Butte Gardens and Arboretum,
    390 Wakara Way, University of
    Utah, Salt Lake City, UT 84108.
    ph. (801) 581-5322
Regional Parks Botanical Garden,
    S. Park Dr. and Wildcat Canyon
    Rd, Tilden Park,
    CA 94708. ph. (510) 841-8732
Santa Barbara Botanic Garden,
    1212 Mission Canyon Road, Santa
    Barbara, CA 93105
    ph. (805) 682-4726
Strybing Arboretum and Botanical
    Gardens, Golden Gate Park, 9th
    Ave. at Lincoln Way,
    San Francisco, CA 94122.
    ph. (415) 661-1316
University of California Botanical
    Garden, Strawberry Canyon and
    Centennial Dr. Berkeley, CA
    94720. ph. (510) 642-3343

CENTRAL UNITED STATES

Bickelhaupt Arboretum,
    340 South 14th St, Clinton,
    IA 52732. ph. (319) 242-1771
Chihuahua Desert Research Institute,
    PO Box 1334,
Texas Hwy. 118, Alpine,
    TX 79831. ph. (915) 837-8370
Eloise Butler Wildflower Garden and
    Bird Sanctuary,
    PO Box 11592, Minneapolis,
    MN 55412.
Houston Arboretum and Nature
    Center, 4501 Woodway Dr,
    Houston, TX 77024.
    ph. (713) 681-8433
Louisiana Nature and Science
    Center, PO Box 870610, New
    Orleans, LA 707187-0610.
    ph. (504) 246-5672
Minnesota Landscape Arboretum,
    Box 39, 3675 Arboretum Dr.
    Chanhassen, MN 55317.
    ph. (612) 443-2460
Nebraska Statewide Arboretum,
    University of Nebraska,
    Lincoln, NE 68583.
    ph. (402) 472-2971
San Antonio Botanical Center, 555
    Funston Place, San Antonio, TX
    78209.ph. (512) 821-5143
The Shaw Arboretum, PO Box 38,
    Gray Summit, MO 63039.
    ph. (314) 742-3512
University of Alabama Arboretum,
    Dept. of Biology,
    PO Box 1927.
    ph. (205) 553-3278
    University, AL 35486
University of Wisconsin Arboretum,
    1207 Seminole Hwy. Madison,
    WI 53711.
    ph. (608) 263-7888
Wild Basin Wilderness Preserve,
    PO Box 13455, Austin,
    TX 78209. ph. (512) 327-7622

EASTERN UNITED STATES

Bayard Cutting Arboretum,
    PO Box 466, Montauk Hwy,
    Oakdale, NY 11769.
    ph. (516) 581-1002
Bok Tower Gardens, Burns Avenue
    and Tower Boulevard, Lake
    Wales, FL 33859-3810.
    ph. (813) 676-1408
Callaway Gardens,
    U.S. Highway 27 South,
    Pine Mountain, GA 31822-2000
    ph. (800) 282-8181
Cheekwood Botanical Gardens, 1200
    Forrest Park Dr, Nashville, TN
    37205. ph. (615) 353-2168
Connecticut Arboretum,
    Connecticut College,
    New London, CT 06320.
    ph. (203) 439-2144

Cornell Plantations,
    One Plantation Rd, Ithaca, NY
    14850. ph. (607) 255-3020
Fernbank Science Center,
    156 Heaton Park Dr, Atlanta, GA
    30307. ph. (404) 378-4311
Garden in the Woods, New England
    Wildflower Society, 180
    Hemenway Rd, Framingham, MA
    01701.ph. (508) 877-7630
Holden Arboretum,
    9500 Sperry Road,
    Mentor, OH 44060
    ph. (216) 946-4400

Matthaei Botanical Gardens,
    University of Michigan,
    1800 North Dixboro Road,
    Ann Arbor, MI 48105
    ph. (313) 998-7061
Memphis Botanical Garden,
    750 Cherry Rd, Memphis, TN
    38117-4699.
    ph. (901) 685-1566
Mount Cuba Center for the Study of
    Piedmont Flora,
    Box 3570, Barley Mill Rd,
    Greenville, DE 19807-0570.
    ph. (302) 239-4244
The North Carolina Arboretum,
    PO Box 6617, Asheville,
    NC 28816. ph. (704) 665-2492
Tower Hill Botanic Garden,
    30 Tower Hill Rd, Boylston,
    MA 01505. ph. (508) 869-6111
The Winkler Botanical Preserve,
    4900 Seminary Rd, Alexandria,
    VA 22311. ph. (703) 578-7888

CANADA

The Arboretum,
    University of Guelph,
    Guelph, Ontario, N1G 2W1.
    ph. (519) 824-4120
Royal Botanical Gardens,
    Box 399, Hamilton, Ontario L8N
    3H8. ph. (416) 527-1158
University of British Columbia
    Botanical Garden,
    6501 NW Marine Dr, Vancouver,
    British Columbia V6T 1W5. ph.
    (604) 228-5828

INDEX *and* GLOSSARY

I n this combined index and glossary, bold page numbers indicate the main reference, and italics indicate illustrations and photographs. Plants are listed under both their common and scientific names.

CAPTIONS

Page 1: Scarlet gilia *(Ipomopsis aggregata)* is a showy wildflower. Hummingbirds and moths are attracted to the nectar, hidden inside its floral tubes.

Page 2: A native wildflower garden will attract a variety of wildlife, particularly butterflies.

Page 3: The colorful flowers of passion flower vines *(Passiflora* spp.) are a magnet to butterflies in summer.

Pages 4–5: Quaking aspen *(Populus tremuloides)* forms beautiful groves, particularly in fall.

Pages 6–7: A tree hollow provides an excellent nesting site for the nocturnal screech owl.

Pages 10–11: There are several species of blanket flower native to North America. *Gaillardia pulchella* makes an attractive garden plant.

Pages 12–13: Every natural garden should contain a quiet spot for people to sit and enjoy nature, and watch the wildlife their garden attracts.

Pages 44–5: Small mammals, such as the eastern chipmunk, are delightful garden visitors, particularly in fall when collecting nuts.

Pages 66-7: Flowering grass-like sedges—such as the star sedge *(Dichromena latifolia)*—provide

convenient landing places for insects such as this young grasshopper.

Pages 100-101: Red-breasted robins light up a bare, wintry garden.

Page 131: Wild buckwheats *(Eriogonum* spp.) and penstemons *(Penstemon* spp.) are colorful flowers for a wildflower garden in a mountainous region.

Page 161: The yellow flowers of desert marigold *(Baileya multiradiata)* light up the landscape.

Page 189: Even in an open prairie garden a few trees can provide useful shade and habitat for birds who prefer shelter to the wide open spaces of the plains.

Page 217: Sweeps of black-eyed Susans *(Rudbeckia hirta)* grow in the meadow gardens of the Northeast region. All plants appreciate a constant water source, such as a pond or stream.

Page 245: A woodland garden is a place to escape from the heat and enjoy a walk along a path through the glades.

Page 273: The two-spotted ladybug beetle is a common garden visitor and is found throughout North America.

ACKNOWLEDGEMENTS

The publishers wish to thank the following people for their assistance in the production of this book:
Gary Fletcher, Stephen Foster, Selena Hand, Diane Harriman, Margaret McPhee, Gail Page, Tracy Tucker.

PICTURE CREDITS

(t = top, b = bottom, l = left, r = right, c = center, bkgr = background.
AA/ES - Animals Animals/Earth Scenes; A = Auscape International Pty
Ltd; APL = Australian Picture Library; BCI = Bruce Coleman Inc, NY;
BCL = Bruce Coleman Limited, UK; BPS = Biological Photo Service;
DJC = DJC & Associates; DRK = DRK Photo; GH = Grant Heilman
Photography, Inc.; NHPA = Natural History Photographic Agency; NS =
Natural Selection; OSF = Oxford Scientific Films; PA = Peter Arnould,
Inc.; PE = Planet Earth Pictures; PI = Positive Images; PN = Photo/Nats,
Inc.; PR = Photo Researchers; TS = Tom Stack & Associates; V = Vireo;
VU = Visuals Unlimited.)

1 D Cavagnaro/DRK. 2 Gay Bumgarner/PN. 3 Dennis Frates/PI.
4-5 Rod Planck/NHPA. 6-7 Karen Bussolini/PI. 8-9 John Shaw/NHPA.
10-11 Leo Meier. 12-13 Saxon Holt. 14t The Granger Collection, NY;
b D Cavagnaro/DRK. 15tl Sonja Bullaty & Angelo Lomeo; tr Stephen J.
Krasemann/DRK; b Jerry Howard/PI. 16t Archive Photos/APL; b Sonja
Bullaty & Angelo Lomeo. 17tl John Gerlach/DRK; tr Mike Read/PE;
b Richard Shiell/AA/ES. 18t Courtesy of Hunt Institute for Botanical
Documentation, Carnegie Mellon University, Pittsburgh, PA. 18-19 Renee
Lynn/PR. 19t Pamela J. Harper; b D Cavagnaro/VU. 20t John Shaw/
NHPA; bl Sonja Bullaty and Angelo Lomeo; br N Schwirtz/BCL. 21tl
Jerry Howard/PI; tr Saxon Holt; b Archive Photos/APL. 22t R & D
Aitkenhead/PI; b Tim Firzharris/Masterfile/Stock Photos. 23 Sonja Bullaty
& Angelo Lomeo. 24 Robert Perron. 24-25 Robert Perron. 25 Mills
Tandy/OSF. 26t Sonja Bullaty & Angelo Lomeo. bl Saxon Holt; br Jerry
Howard/PI. 28 Harry Haralambou/PI; tr Jerry Howard/PI. 29t Jerry
Howard/PI; bl Courtesy of Biltmore Estate, Asheville, North Carolina;
br & bkgr Robert Perron. 30t John Neubauer; b Sonja Bullaty & Angelo
Lomeo. 30-31 Gay Bumgarner/PN. 31t John Neubauer; b Jerry Howard/
PI. 32t Michael P Gadomski/BCI; r Charles Mann; b D & I MacDonald-
NMR/Envision. 33l Richard Shiell/AA/ES; tr Robert W Ginn/Envision;
b Jerry Howard/PI. 34t Michael Habicht/AA/ES; b Dan
Suzio/PR. 35tl Margaret Hensel/PI; tr Carole Ottesen; b D Cavagnaro/VU.
36t Sonja Bullaty & Angelo Lomeo; r J A L Cooke, OSF/ AA/ES. 37t
Karen Bussolini/PI; r Jerry Howard/PI. 38t Bruce M Herman/PR. 38t Lee
Lockwood/PI; b Jerry Howard/PI. 39l Robert A Lubeck/AA/ES; tr
Margaret Hensel/PI; c & b Jerry Howard/PI. 40t Scala/ Art Resource,
NY; b Brian Kenney/Natural Selection. 41t Stephen J Krasemann/NHPA;
r Margaret Hensel/PI; b John Shaw/NHPA. 42t Saxon Holt; c Charles
Campbell/Westlight/APL. 42-43 Francois Gohier/A. 44t Pat O'Hara;
c Tom Bean/DRK; r Stephen G Maka/PN. 44-45 Daniel J Cox/DJC.
46-47 Dan Griggs/NHPA; b Jerry Howard/PI. 47 Saxon Holt. 48t Patti
Murray/AA/ES; b Charlie Palek/AA/ES. 49l Les Campbell/PI; tr Sonja
Bullaty and Angelo Lomeo; b Saxon Holt. 50t Martin Mills/PI; r Dick
Canby/PI; b Elvan Habicht/AA/ES. 51t Saxon Holt; b John Gerlach/DRK.
52t Tom & Pat Leeson/PR; b Jerry Howard/PI. 53t Charles Mann;
c Doug Wechsler/AA/ES; b Daniel J Cox/DJC. 54 Pamela J Harper. 55t
Rich Kirchner/NHPA; tr Leonard Lee Rue IV/BCI; cr Ken Cole/AA/ES;
bl S Nielsen/DRK; br Doug Wechsler/AA/ES. 56t Wayne Lankinen/DRK;
r Link/VU. 57tl Mary Clay/PE; tr David McDonald; bl Carole Ottesen;
br Jon Mark Stewart/BPS. 58t Jerry Howard/PI; r John Neubauer;
b Margaret Hensel/PI. 59t Kathy Merrifield/PR; l W J Weber/VU; b D
Cavagnaro. 60t Stephen Dalton/NHPA; r Charles Mann; b Dennis Frates/
PI. 61t Virginia Twinam-Smith/PN; l Jerry Howard/PI; b Wayne
Lankinen/PI. 62bl John Neubauer; r D Cavagnaro. 62-63 Saxon Holt.
63 John Shaw/NHPA. 64t The Granger Collection, NY; r Jerry Howard/
Positive Images. 65tl Patti Murray/AA/ES; tr Priscilla Connell/Envision;
br & l Jerry Howard/PI. 66-67 Leo Meier. 68t S Dalton/NHPA; b Jacob
Mosser III/PI. 69t Margaret Hensel/PI; c John Shaw/NHPA; b John
Shaw/A. 70t Brian Kenney/PE; r R J Erwin/DRK; b Paul Rezendes/PI.
71 John Gerlach/AA/ES. 72t Brian Kenney/NS; b Joe McDonald/NS.
73t Brian Kenney/NS; bl Daniel E Wray/NS; br Brian Kenney/NS.
74t Leo Meier; bl Patti Murray/AA/ES; br A Carrara/Jacana/A. 75t Thomas
A Schneider/ NS; b E R Degginger/AA/ES. 76t E R Degginger/ PR;
b David M Dennis/TS. 77tl Reed Williams/AA/ES; tr Richard Shiell/AA/
ES; b Syd Greenberg/PR. 78t George Sanker/DRK; bl Milton Rand/TS;
br John Gerlach/DRK. 79t Dwight R Kuhn/BCI; b S. Nielsen/DRK.
80t Ken Highfill/PR; r J & L Waldman/BCI; b John C Parker/PI. 81t
Laura Riley/ BCI; b S Nielsen/BCI. 82tl The Granger Collection, NY; tr
D Cavagnaro/DRK; b W Greene/V. 83 Robert A Tyrrell/OSF/AA/ES.
84bl Robert A Tyrrell; br C H Greenwalt/V. 85tl HummZinger by
Aspects/photo C Fogle; tr Robert A Tyrrell/OSF/AA/ES; b The Granger
Collection, NY. 86t John Gerlach/DRK; b Les Campbell/PI. 87tl Stephen
J Krasemann/DRK; tr John Parker/PI; bl Bob & Clara Calhoun/BCL;
br Stephen J Krasemann/DRK. 88t Dogwood and Mockingbird by Mark
Catesby from the National History of Carolina (1730-48)/Lindley Library,
RHS, London/The Bridgeman Art Library, London; c Paul Rezendes/PI;
bl John Cancalosi/DRK; br Barbara Gerlach/DRK. 89t John Gerlach/
DRK; c John Shaw/NHPA; b Hal H Harrison/GH. 90t Joe McDonald/
BCI; b Norman R Lightfoot/PR. 91tl Daniel J Cox/DJC; tr ER Degginger/
BCI; b Fred Whitehead/AA/ES. 92tl & r John Serrao/PR; c C W Perkins/
AA/ES; b Daniel J Cox/DJC. 93t Alfred B Thomas/AA /ES; c Jeff Foott/
DRK. 94t Steven M Rollman; r E R Degginger/DRK; b James R
Fisher/DRK. 95t Merlin D Tuttle/Bat Conservation International Inc;
b Brian Kenney/PE. 96t Bob Gurr/DRK; b Leonard Lee Rue III/BCI.
96-97 Mella Panzella/AA/ES. 97t Jerry Howard/PI. 98bkgr Les Campbell/

PI; tr Colin McRae; bl Oliver Strewe; br John Gerlach/DRK. 99t Archive
Photos/APL; r Reproduced by permission of the American Museum in
Britain, Bath. 100-101 Daniel J Cox/DJC. 102t Courtesy of Hunt Institute
for Botanical Documentation, Carnegie Mellon University, Pittsburgh, PA;
b D Cavagnaro/DRK. 102-103 Ken Druse, The Natural Garden.
104-105 Saxon Holt. 108 Harry Smith Collection. 109 Ken Druse/The
Natural Garden. 110 Harry Smith Collection. 111 Saxon Holt. 112 Charles
Mann. 113 Dr Glenn Keator. 114 Eric Crichton/ BCL. 115 Saxon Holt.
116 Eric Crichton/BCL. 117 Saxon Holt. 118 C. Colston Burrell. *
119 Roger Raiche. 120 Carl May/BPS. 121 Saxon Holt. 122 Bruce A
Macdonald/AA/ES. 123 Robert C. Perry. 124 David McDonald. 125 Carl
May/BPS. 126 Jack Wilburn/AA/ES. 127 Saxon Holt. 128 D Cavagnaro.
129 David Middleton/NHPA. 130 John Shaw/TS. 131 Dr Glenn Keator.
132l Hal Benzl/VU; r D Cavagnaro. 133 Saxon Holt. 136 Charles Mann.
137 George F Godfrey/AA/ES. 138 Charles Mann. 139 Stephen J
Krasemann/DRK. 140 Saxon Holt. 141 Charles Mann. 142 David A
Ponton/PE. 143 Tom Bean/DRK. 144 Barbara J. Miller/BPS. 145 Jim
Knopf. 146 Charles Mann. 147 W D Bransford/ National Wildflower
Research Center. 148 Milton Rand/TS. 149 Valorie Hodgson/PN.
150 David A Ponton/ PE. 151 Kahout Productions/Root Resources.
152 Larry Ulrich/ DRK. 153 Charles Mann. 154 Andy Wasowski.
155 Patti Murray/AA/ES. 156 Breck P Kent/AA/ES. 157 Willard
Luce/AA/ES. 158 David Middleton/NHPA. 159 Richard Shiell/AA/ES.
160l Saxon Holt; r Robert E Heapes. 161 Charles Mann. 164 Larry Ulrich/
DRK. 165 Rod Planck/ NHPA. 166 Doug Sokell/TS. 167 Kahout
Productions/ Root Resources. 168 Patti Murray/AA/ES. 169 Leo Meier.
170-171 Andy Wasowski. 172 Charles Mann. 173 Saxon Holt.
174 Andy Wasowski. 175 John Cancalosi/DRK. 176 Ken Druse, The
Natural Garden. 177 Mickey Gibson/AA/ES. 178 Jon Mark Stewart/ BPS.
179 Charles Mann. 180 Andy Wasowski. 181 Stan Osolinski/OSF.
182 Scooter Cheatham. 183 Charles Mann. 184 Saxon Holt. 185 Andy
Wasowski. 186 Gregory G Dimijian/PR. 187 Francois Gohier/A.
188l Charles Mann; r W D Bransford/National Wildflower Research
Center. 189 Saxon Holt. 192 Charles Mann. 193 Stephen Trimble/DRK.
194 Susan Gibler/TS. 195 Zig Leszczynski/AA/ES. 196 John Lemker/
AA/ES. 197 Ken Druse/ The Natural Garden. 198-200 Charles Mann.
201 Ken Druse/The Natural Garden. 202 Leo Meier. 203-4 Charles Mann.
205 Doug Wechsler/AA/ES. 206 Pam Spaulding/PI. 207 Robert E
Heapes. 208 David Stone/PN. 209 Jeff March/PN. 210 Charles Mann.
211 David McDonald. 212 E R Hasselkus/Arboretum, University of
Wisconsin, Madison. 213 Wendy Shatil & Bob Rozinski/TS. 214 Pat
Kennedy/NS. 215 Harry Smith Collection. 216l D Cavagnaro; r Leo
Meier. 217 Sonja Bullaty & Angelo Lomeo. 220 David Stone/PN.
221 Jerry Howard/PI. 222 T E Degginger/ AA/ES. 223 Rod Planck/
NHPA. 224 David Stone/ PN. 225 Mary Clay/ PN. 226 Ann Reilly/PN.
227 Ken Druse/The Natural Garden. 228 Jeff Lepore/PR.
229 Stephen G Maka/PN. 230 Michael P Gadomski/AA/ES. 231 Paul
Rezendes/PI. 232-233 Ken Druse/ The Natural Garden. 234 Les Campbell/
PI. 235 Patti Murray/AA/ES. 236 Pam Spaulding/PI. 237 Walter H
Hodge/PA. 238 Robert P Carr/BCI. 239 Jerry Howard/PI. 240 Larry
Lefever/GH. 241 David McDonald. 242 Pamela J Harper. 243 Jerry
Howard/PI. 244l Jerry Howard/PI; r William J Weber/VU.
245 Saxon Holt. 248 Wendell Metzen/BCI. 249-251 Ken Druse, The
Natural Garden. 252 Robert E Lyons/PN. 253-255 Pamela J Harper.
256 Jerry Howard/PI. 257 Bob Hyland/Brooklyn Botanic Garden.
258 Stan Osolinski/OSF. 259 D Cavagnaro. 260 John Shaw/NHPA.
261 Larry Lefever/GH. 262 Michael P Gadomski/PR. 263 Eric Crichton/
BCL. 264 David McDonald. 265 Andy Wasowski. 266 Lucy Jones/VU.
267 Jack Dermid/BCL. 268 Tyler Gearhardt/Envision. 269 Deni Brown/
OSF. 270 Harry Smith Collection. 271 Dr Glenn Keator. 272l Steve
Solum/BCI; r Andy & Sally Wasowski. 273 S Neilsen/DRK.

JACKET CREDITS

Front: Wayne Lankinen/DRK; Stephen Dalton/NHPA; Brian Kenney/PE
Back: Stephen Dalton/NHPA **Front flap:** Jerry Howard/PI; Saxon Holt
Back flap: Paul Rezendes/PI **Spine:** Patti Murray/AA/ES

ILLUSTRATION CREDITS

Simon Darby 16.
Frank Knight 83.
Angela Lober 46, 110, 111, 113, 121, 123, 129, 131, 153, 178, 180, 185,
210, 249, 250, 251, 252, 255, 256, 257, 263, 264, 266.
Rob Mancini 62, 71, 112, 114, 115, 119, 122, 125, 127, 128, 130, 136,
138, 144, 146, 152, 156, 157, 158, 165, 166, 168, 171, 176, 177, 179, 193,
194, 198, 207, 208, 212, 213, 215, 222, 226, 229, 230, 233, 235, 237, 242,
248, 253, 254, 258, 260, 261, 262, 265, 267, 268.
Nicola Oram 93, 108, 126, 137, 139, 140, 141, 142, 143, 145, 147, 149,
150, 151, 154, 164, 167, 170, 172, 181, 182, 183, 184, 187, 195, 196, 197,
199, 203, 204, 206, 209, 214, 220, 221, 223, 224, 225, 227, 231, 232, 236,
240, 241, 243, 269.
Ngaire Sales 55, 106, 107, 132, 133, 162, 163, 190, 191, 218, 219, 246, 247.
Genevieve Wallace 109, 116, 117, 118, 120, 124, 148, 155, 159, 169,
173, 174, 175, 186, 192, 200, 201, 202, 205, 211, 228, 234, 238, 239, 259,
270, 271 and resources directory.
David Wood 23, 24, 27.